MIXING RACES

JOHNS HOPKINS
INTRODUCTORY STUDIES
IN THE HISTORY
OF SCIENCE

Mott T. Greene
and Sharon Kingsland
Series Editors

Mixing Races

From Scientific Racism to Modern Evolutionary Ideas

Paul Lawrence Farber

THE JOHNS HOPKINS UNIVERSITY PRESS

BALTIMORE

© 2011 The Johns Hopkins University Press
All rights reserved. Published 2011
Printed in the United States of America on acid-free paper
9 8 7 6 5 4 3 2 1

The Johns Hopkins University Press
2715 North Charles Street
Baltimore, Maryland 21218-4363
www.press.jhu.edu

Library of Congress Cataloging-in-Publication Data

Farber, Paul Lawrence, 1944–
    Mixing races : from scientific racism to modern evolutionary ideas /
Paul Lawrence Farber.
        p. cm. — (John Hopkins introductory studies in the history of science)
    Includes bibliographical references and index.
    ISBN-13: 978-0-8018-9812-9 (hbk. : alk. paper)
    ISBN-10: 0-8018-9812-9 (hbk. : alk. paper)
    ISBN-13: 978-0-8018-9813-6 (pbk. : alk. paper)
    ISBN-10: 0-8018-9813-7 (pbk. : alk. paper)
    1. United States—Race relations—History—20th century.
2. Miscegenation—United States—History—20th century. 3. Interracial
marriage—United States—History—20th century. 4. Interpersonal relationships
and culture—United States—History—20th century. 5. Human population
genetics—Social aspects—United States. 6. Science—Social aspects—United
States. I. Title.
    E185.62.F37 2011
    305.800973—dc22   2010017556

A catalog record for this book is available from the British Library.

The illustration on page 47 is courtesy of the United States Holocaust Memorial
Museum. The views or opinions expressed in this book, and the context in which
the image is used, do not necessarily reflect the views or policy of, nor imply
approval or endorsement by, the United States Holocaust Memorial Museum.

*Special discounts are available for bulk purchases of this book. For more information,
please contact Special Sales at 410-516-6936 or specialsales@press.jhu.edu.*

The Johns Hopkins University Press uses environmentally friendly book materials, including recycled text paper that is composed of at least 30 percent post-
consumer waste, whenever possible. All of our book papers are acid-free, and our
jackets and covers are printed on paper with recycled content.

In memory of Anita Schapiro Michaels
and Helen Shapiro Farber

Contents

Acknowledgments

This book began rather indirectly. Roughly ten years ago, I visited our twins, Channah and Benjamin, at Cornell, where they were undergraduates. While in Ithaca, I took the opportunity to do some work in the library on a project about late nineteenth-century French ideas on evolutionary ethics. In the rare book room, I happened across a report by Jean de Lanessan on the condition of the French colonies. Lanessan noted in his discussion of Southeast Asia that the children of colonials and natives seemed better suited to the environment than either parent stock. I was quite struck by the remark and thought that perhaps I had chanced upon a previously overlooked school of thought that viewed racial mixture in a positive light—most of the scientific literature of the late nineteenth and early twentieth centuries viewed the offspring of interracial couples as degenerate. For several months, I attempted to run down similar French statements or studies but in the end concluded that Lanessan's opinion was not part of an obscure French medical tradition; it was more likely just an offhand comment.

But it got me thinking. I wondered when attitudes on race mixing had shifted and what had led to the change. The topic resonated with me, partly because of a vivid memory I had of my undergraduate days, when the dean of women was rumored to have broken up a mixed-race couple (by notifying the parents and having the students removed from school). The event had been all the more shocking coming as it did during the early sixties, when the subject of civil rights dominated much of the campus. Although I hadn't planned to do a project on race mixing, the issue continued to intrigue me, and I soon found myself embarked on the research project that has resulted in this book.

As with any project that spans roughly a decade, I have been helped by many people, too many to list. I would be remiss, however, not to mention some who have been particularly important for me.

Vreneli Farber, who has patiently provided a sounding board and moral support for all my past work, became actively involved in this one, and it was her help in the library and archives at the University of Pittsburgh that provided critical pieces for my research. My friendship with Ernie Graves in the sixties taught me more about race, and about myself, than I can ever adequately thank him for. Lucy Correnti and Jim Spruill generously took time to talk with me about their difficult experiences in the sixties and gave me insight into events that I witnessed but did not fully understand at the time. Russ Barnes, who has been a friend since fourth grade, was a valuable source of information about life in western Pennsylvania in the late 1950s and early '6os.

I am indebted to many scholars, and my notes reflect some of their published works. Early on, the scholarship of Paul Spickard and Werner Sollors helped me get oriented to the subject. Conversations with and shared materials from colleagues have been as important as published works: Will Provine sent me the manuscript of an unfinished book on the history of attitudes to race mixing; Mark Largent spent hours re-educating me about Davenport; Gar Allen spent even more hours plying me with Scotch to remind me how much I loved doing history of science; Andrew Valls helped expand my concept of race; Dee Baer stressed that I get my science correct; Hamilton Cravens helped me understand the historical context of this project; Melinda Gormley helped me better appreciate Dobzhansky's intellectual life at Columbia and his association with L. C. Dunn; Mott Greene spent hours saving me from myself by helping me rethink and reshape an earlier manuscript; and Kristin Johnson also spent a lot of time discussing the many dimensions of the subject and earlier versions of this book.

Librarians and archivists, who most certainly must occupy some special circle in Paradise, have been invaluable. Cliff Mead of Oregon State University and Marianne Kasica of the University of Pittsburgh have been especially important. Libraries that made it possible for me to do the research necessary for this book were the British Library, the Valley Library at Oregon State University, the Hellman Library at University of Pittsburgh, the Herman B. Wells Library at Indiana University, the Bancroft Library at University of California, Berkeley, and the Bernard Becker Library at Washington University's School of Medicine. Oregon State University Archives, University of Pittsburgh Archives, and the American Philosophical Society Library Archives made valuable manuscript material and rare printed matter available to me.

Bob Brugger, of the Johns Hopkins University Press, encouraged me throughout the process of writing, even when he was trashing my manuscript, and I owe him a special debt of gratitude for all the help he has given me over the years.

And finally, I am deeply aware of the valuable friendship and intellectual stimulation my colleagues in the History Department at Oregon State have provided me all these years. They not only made this project possible, but also made it fun.

MIXING RACES

Introduction

In the 1920s and thirties, it was not unusual for medical experts to caution the public that a mixed marriage ran the risk of producing biologically dysfunctional offspring. The commonly accepted opinion at the time stipulated that since races differed so strikingly from one another, the genetic material coming from each parent might not align properly in the children, and therefore a "chaotic constitution" could result. This biological warning reinforced already strong social taboos that permeated the culture and found expression in state laws (and earlier colonial laws) outlawing interracial marriage.

By the 1960s, however, the idea that race mixing might be biologically injurious, or that the state had the right to prohibit such interracial unions, had met serious challenge. The U.S. Supreme Court decision in *Loving v. Virginia* (1967) marked a turning point in American racial history when it ruled unconstitutional—a violation of "equal protection"—for states to outlaw marriages between individuals of different races. The number of interracial marriages has increased since then, decade by decade. If still a minority, mixed-race families nevertheless are a part of America's social landscape.

During the first decade of the twenty-first century, celebrities of mixed-race ancestry attracted increased media coverage of interracial marriage. Tiger Woods declared on the *Oprah Winfrey Show* that he was not "black" but was of "mixed race," and during the presidential campaign of 2008, Barack Obama frequently mentioned his white mother (from Kansas) and black father (from Kenya).[1] Although the national and international press have treated Obama as the first African American president, they have also used his election as an

---

1. In this work, various terms will be used to designate popular conceptions of race: White, Black, African American, Asian, Indian, Native American, Negro, Colored, Caucasian, and so forth. Usually, the term used is appropriate to the historical context, but in several places, more than one term would work. The terms *white* and *black* have generally not been capitalized except in those cases where they are used specifically as proper nouns designating classifications of races rather than adjectives implying skin color. The term *race* itself is typically used in its everyday popular sense and does not imply any scientific meaning.

opportunity to bring to the fore a discussion of race mixing and children of mixed-race families. There has been no suggestion that individuals like Obama might suffer from a "chaotic constitution" or be biologically disadvantaged. Rather, journalists have focused more on questions about identity. For example, why should the child of black and white parents generally be considered African American? How many generations of mixing with white partners are necessary to "remove" the offspring from the category of black? (Unlike the search for aristocratic ancestry, blackness has too often been considered a "taint" that can be diluted over several generations, rather than a benefit that bestows status, even when removed by a century.)

Increasingly, the public recognizes the inadequacy of current racial categories to capture the identity of mixed-race children. In the 2000 federal census, respondents for the first time could indicate their racial identity by marking more than one of five categories (White; Black or African American; American Indian or Alaska Native; Asian; and Native Hawaiian or Other Pacific Islander).

Race has been one of the central social issues in the history of the United States, and ideas about race mixing have been touchstones for gauging how Americans think about race. Much of the justification for the policy of segregated schools in the South, to take one example, rested on the fear that social integration of white and Negro students would lead to romantic integration and, with that, miscegenation.

During the past few decades, historians have created a large literature on race and, in particular, on the scientific racism that Americans used for many years to legitimate discrimination and the separation of races. Scholars described how the intellectual world rejected scientific racism after the Second World War, and they noted the various factors that played into that transformation. The larger society more slowly shifted its opinion. Racism did not disappear in the late 1940s; however, some of the most important justifications that had been used for discriminatory policies were discredited. By the 1960s, the civil rights movement could draw on the altered attitude toward race within the scientific and intellectual communities to advance its cause and to further reduce the ill effects of racism.

This book explores the changing American view on race mixing in the twentieth century, particularly the critical years between roughly 1940 and 1970. During this time, U.S. society experienced a set of radical transformations culminating in the "Sixties Revolution." Attitudes about race mixing were one of the major cultural alterations that occurred. The first part of this study asks how and why the scientific community rejected the racial theories that had legitimated the notion that race mixing could be biologically deleterious. Historians have described this shift primarily as a result of social factors, especially the revulsion against Nazi atrocities and their racial propaganda, as well as the xenophobic nationalism throughout the Western world that had contributed to the origins of two devastating world wars. Cultural anthropologists also played an important role in changing attitudes during and after the Second World War by arguing that there were no significant biological differences among European "racial groups," such as the French and the Germans. Instead, these populations were described as mixtures of many stocks, none of them "pure." This more liberal social science viewpoint undermined the eugenic literature that at the turn of the century had disparaged immigrants to the United States from southern and eastern Europe. The new ideas about European "races" had the unintended consequence of loosening the divisions between the "major" races: Caucasian, Negro, and Asian, which anthropologists also came to view as broadly mixed populations.

Historians have failed, however, to appreciate fully how changes in the natural sciences contributed significantly in altering discussion of race and race mixing. Although no new surprising *empirical* investigations were involved, deeper *theoretical* changes had a profound impact. Science may have been part of the problem in justifying the erection of social barriers based on race, but science also played a part in dismantling them.

How did changes in scientific perception translate into social attitudes and actions? Although the realization that race mixing was not biologically dangerous did not lead to a sudden increase in miscegenation, changes in ideas did create significant social and intellectual tensions, especially in the 1960s on university campuses, and for good reason: universities were social laboratories, where new ideas circulated and where young people of different races and of marriageable age found themselves in close proximity. Students and faculty were often ahead of administrators (who felt the pressures of alumni and parents). The tensions generated on campus reliably indicated deeper shifts in the social fabric.

Campus life in the 1960s had a profound effect on U.S. culture. That importance was partly because media coverage of campus disturbances brought them to the public eye, and partly because the newly expanded universities reflected changes the country was undergoing at the time. The country had been through a decade of "returning to normal" after the Second World War, and a growing prosperity gave a larger percentage of the population access to higher education. Women and minorities had taken part in the war effort, and their legacy expanded the aspirations of a new generation. The sixties became a watershed decade, and events on campus had important ramifications throughout society at large.

A set of events that occurred at the University of Pittsburgh in the early 1960s are an entry point to this story. Situated in a progressive city with a large black population, the university experienced a restructuring of its student population in this decade. Part of that change involved greater inclusion of African American students and foreign students of color. New social problems arose as a consequence of the diversifying student body, and those problems reflected events happening across the country on other campuses and in the larger society.

Another reason for beginning in Pittsburgh is that my wife and I were students there at the time, and we experienced some of these pivotal events firsthand. Our diaries, college notes, letters, and conversations with friends who were there provide an unusual set of resources. If portions of the narrative sound autobiographical, it is not by accident; however, historical sources document claims that are made.

1 A Mixed-Race Couple in the 1960s

They were easy to spot: eating by themselves in the cafeteria; moving through the quad of Holland Hall; or sitting in the University of Pittsburgh's student union lounge. He was tall, thin, and black, with a white cane, and she short, slightly chubby, and white, in a wheelchair. Both were nicely dressed. The boy generally wore a sports jacket and a dress shirt, buttoned at the collar. The girl favored cotton dresses in pastel colors. They looked like they had families who cared about them. When crossing the quad, or moving from the lounge to the cafeteria, she would carry his cane while he pushed her wheelchair. She was his eyes, and he was her legs. They laughed a lot: uninhibited laughter, not too loud, not calling attention to itself, but genuine, warm. Two young people who, when seen separately changing classes, might have elicited a flicker of sympathy, together formed an enviable couple that neutralized the daily hassles of being "handicapped" on an urban campus.

"Mixed couples" in the early 1960s stood out. Less so perhaps in Pittsburgh than in Montgomery, Alabama, but nonetheless, on that largely white campus, a biracial couple would have been noticed by other students and by faculty. The pair, perhaps because their individual disabilities defined them more than their skin colors, might not have struck onlookers as a mixed couple. They became partly de-sexed because of their disabilities and therefore did not shout "black man" and "white woman." Certainly, their behavior worked to downplay a romantic image. When they were together in the union lounge, they weren't clutched in each other's arms with hormone-glazed eyes like the other couples there. They sat facing one another with radiant smiles. To the casual observer, their friendship may have looked platonic. They were a threat nonetheless, although not as much as another mixed couple who surfaced later in the year: she a beautiful and slender black female, and he an athletic, handsome white male. These two made no effort to hide their physical attraction for each other, and the romantic dimension of their relationship was obvious.

Both couples attracted the attention of the university administration. In the spring of 1964, students were shocked by a widespread rumor that the dean of women had contacted the parents of the mixed-race couples and had arranged for the students to leave the university. She found their presence intolerable and believed that the students would be better off if their misalliance were terminated. Part of her responsibility as dean was the well-being of her charges, and she took this *in loco parentis* role seriously.

To students, the rumor highlighted how much the university administration was out of touch with society. Newspapers and public relations efforts presented Pittsburgh as a progressive city. It had a large black population and extensive ethnic neighborhoods. Travelers had called it the smoky city before and during the war—so bad that the smog and smoke created near-nighttime conditions at mid-morning—but the mayor in the 1950s had worked to pass stringent laws that cleaned up its notoriously polluted air. The city was home to the *Pittsburgh Courier*, a widely respected black newspaper. Blacks and whites mixed freely in jazz bars like the Crawford Grill, August Wilson's hangout in the Hill District. At the University of Pittsburgh, many professors encouraged students to embrace the promise of the American dream, which included the vision of a color-blind society, one where individuals aspired to fulfill their potential and those of diverse backgrounds mixed freely.

On Campus in the 1960s

The first half of the 1960s was an exciting time to be a college student. Universities enjoyed a period of extended growth that had started after the war with the GI bill and continued with the enrollment of the children of those GIs—the cusp of the baby boomers. State budgets supported new building on campuses. In response to the Cold War, students could receive full support from government agencies by majoring in Russian or Chinese, and professors with expertise in Middle Eastern political science or Soviet geography regularly shuttled back and forth between their home institutions and Washington as highly paid consultants. Even somewhat esoteric fields like philosophy of science received funding from NASA (for research on the philosophy of space and time). Although the percentage of students receiving a postsecondary education was still small compared to what it would become in the eighties and nineties, a university degree was no longer confined primarily to the children of an elite group of professionals and American "upper crust." Growth meant change, and change charged the atmosphere.

What a freshman first noticed on most college campuses, however, even on an urban one like the University of Pittsburgh, was the vibrant fraternity and sorority scene. The majority of students knew the Greek alphabet, if not its grammar, and the most visible social life of the university centered on the many Greek houses that dotted the perimeter of the campus. They were, however, living on borrowed time. The conformity, the ritual, and the blatant materialism that characterized those groups had seen their halcyon days in the fifties. The social order, if not the social scene, was in motion, and like geological change, which often moves at an imperceptibly slow pace but can build to enormous pressure, it was altering the landscape of American society and American campuses.

Along with ΣAE and $\Delta\Delta\Delta$ mixers, new students found themselves confronted with stridently vocal and highly visible political groups like SANE (Committee for a SANE Nuclear Policy) and the more radical SDS (Students for a Democratic Society). The Cuban missile crisis of 1962 convinced many young people that the stability and security they had taken for granted was an illusion. It shouldn't have been a surprise. After all, a steady rumbling had persisted throughout their childhood: grade school air raid drills, an unresolved Korean war, images on the television of the atomic and hydrogen bombs and of Senator Joseph McCarthy's public paranoia over the infiltration of communists throughout America.

When parents, even those who had been left-leaning ("pink") in the 1920s and thirties, expressed concern about their children's interest in SDS, it was with memories of their innocent, but naïve, friends who had later been blacklisted during the McCarthy period because of their membership in allegedly subversive organizations and had therefore been made unemployable. The sixties generation, however, was not seeking a return to normalcy; they were the beneficiaries of that movement. They had been raised with an idealistic, patriotic vision, designed to counter the menace of creeping communism and its threat to Freedom and Democracy, but they had absorbed those lessons literally. To them, the compromises that an older generation had learned to live with felt like a betrayal of those ideals for which our country had fought so hard. Why shouldn't they join whatever political group interested them? They were students who wanted to be engaged in the world; many of them would join the newly founded Peace Corps.

Communism, atomic warfare, and third world aid were not the only issues that energized students. What would be called the civil rights movement emerged as a central driving issue on many American campuses in the early sixties. The court-mandated integration of southern universities set off a chain reaction of events on campuses across the country, especially on northern campuses. The earlier public school integration that followed the Supreme Court decision *Brown v. Board of Education* (1954) had sensitized the public to the issue of civil rights, but the TV and press coverage of the dramatic events surrounding the integration of the University of Mississippi and other southern schools of higher education struck a chord with undergraduates in the North. Students engaged in freedom rides, voter registration campaigns, and protest marches to demonstrate their commitment to full civil liberties for *all* Americans. Students who were ready to board a bus to join voter registration drives in the South, and face the hostility of local crowds, were not students who wanted to be coddled.

What would a student at this time think about issues of race, and more specifically, race mixing? Let us stay, for the moment, in Pittsburgh, where this story began. In the early sixties, Pitt described itself as semipublic—that is, a private school with some state support (and hence some public obligations). It drew the majority of its students from Pennsylvania. Edward Litchfield, then chancellor, had been involved in the reconstruction of Germany after the Second World War, and in 1956, with the help of some philanthropic families (Falk, Scaife, Mellon), he had undertaken the academic expansion of the university, converting it from a less-than-distinguished commuter school into a first-rate institution of higher learning. By the early 1960s, he had made impressive gains by raiding several elite institutions and luring a cadre of distinguished professors to fill a dozen endowed chairs. They, in turn, attracted other highly competent faculty and initiated a recruitment phase that helped transform the university. Litchfield was highly visible—walking along the corridors of the Cathedral of Learning (the 535-foot, 42-story Gothic skyscraper that is at the heart of the campus) with a martini glass (on a dry campus!), or waving to the undergraduates at a football game from his box, with Zsa Zsa Gabor on his arm. Student enrollment came to include significant numbers from out of state and a noticeable international student presence. Undergraduate and graduate students from India, the Congo, the British Isles, and the Middle East pursued degrees in various subjects, and they took part in the social life of the campus. Students, therefore, often found themselves rooming with someone from a quite different background. It would not be unusual to have in Schenley Hall (a dorm recently converted from a formerly deluxe hotel) a white, Jewish, premed, middle-class boy rooming with a Southern black (Colored or Negro were the terms back then) engineering major. A number of southern black students had chosen to attend school in the North because they wanted to escape the social constraints of the segregationist South. The North promised more opportunities.

Western Pennsylvania in the early sixties was hardly free from prejudice, however. For example, a local high school history teacher, upon being informed that a former student of his had a black roommate, wrote his student a caustic letter, asking, "In what field is your roommate majoring? It occurs to me that there are many potentialities among his race for him to explore; with their high rates of illiteracy, unemployment, illegitimacy, venereal disease, alcoholism, 'reliefism,' and so on ad infinitum, the American negroes afford him ample scope in which to exercise his abilities."[1]

But by and large, the university was more liberal. The rhetoric of the Cold War in the sixties emphasized the openness of American society: the freedoms of expression, religious belief, and association. The United States was a country of *liberty,* in contrast to those communist states trapped behind the Iron Curtain. Although only the bravest of professors openly discussed Marx, historians, sociologists, and anthropologists lectured on social class and argued that class did not imply a hereditary order, that individuals could move up in social class—the American Dream was open to all. And yet, for parts of the administration, race mixing remained beyond the pale.

The "administration," of course, was a somewhat nebulous concept. Undergraduates did not have much of a sense of what constituted the administration, or who did what. At Pitt, students were mostly aware of the charismatic chancellor. Noticeable other parts of the administration consisted of academic advisers, department chairs, and, of course, the endless number of (mostly) unhelpful staff in the registrar and business offices. Students were also aware of the dean of men and the dean of women. The deans interacted with students in many ways, much of it in their *in loco parentis* role. Pitt had a policy of no alcohol on campus, and there were dress codes and curfews for women. Any of a number of wayward behaviors was likely to land you in front of one of the two deans.

Of all the administrators, the dean of women, Helen Pool Rush, and her assistant, Savina Skewis, were those most likely to catch the attention of students. Although men outnumbered women at Pitt, the restrictions on women were greater, and therefore the reach of their dean's office was further. The dean of women had a suite of offices on the twelfth floor of the Cathedral of Learning. For years the assistant dean was a mentor to a select group of coeds (regularly serving them tea) who in their final year were mentors to incoming freshmen women. She made sure that these young women adhered to the highest moral principles. Aspects of the recruitment policy reflected her moralistic tone. Proscribed behavior by any freshman or sophomore female students, such as staying out after curfew, was to be reported to the assistant dean so she could swing into action: calling the miscreant into her office and

1. Letter from Con Costolo, Oct. 8, 1961, Farber Papers, Special Collections, Oregon State University Valley Library.

In loco parentis

Until the mid-1960s, undergraduates had to adhere to a set of rules that governed their private lives. Women in particular had considerable restrictions. Unmarried female undergraduates under 21 were often required to live in dorms (which were single sex), or with relatives (not in any other private homes or apartments). Generally, colleges had curfews for women. Coeds, as female students who attended coeducational schools were called then, had to sign out of the dormitory if they were out after 7:30 P.M. and had to have written permission to travel out of town. A dress code was common: women could not wear slacks to class, in the library, or in the cafeteria. Campuses were mostly dry, and drinking was strictly forbidden in the dormitories. (Smoking was OK.) Infrac-

tions of the rules could lead to suspension. Universities felt it was their responsibility to expel students who were thought to be "immoral." That could cover a wide range of sins, especially sexual "laxness." Schools reserved the right to prohibit non-university organizations from demonstrating or soliciting donations.

In addition to formal rules, expectations existed about what was acceptable in behavior and attire. Coed handbooks, from which the accompanying illustration comes, conveyed some of these expectations and covered a range of topics, from which social occasions called for a hat and gloves to proper phone etiquette with boys. Some of the advice had to do with adjusting to college life and working effectively there. But much of it reflected the watchful eye of the Dean of Women's Office, whose staff members believed they were games will be 1:30 a.m.

f. Automatic extended closing hours will be in effect after major all-school dances ending at midnight or after. These are Homecoming, Junior-Senior Prom, Sophomore Cotillion and Mortar Board Ball. Time is extended to 2:00 a.m.

g. Closing hours after Thanksgiving vacation and the night before the first day of registration after Christmas and spring vacation shall be 12:00.

h. Exceptions to closing hours will be announced and published through the Barometer prior to the event.

i. Individual exception to opening or closing hours is granted by the Student Board of Reference or the chairman of the board. Groups, classes or individuals representing the college requiring later or earlier hours should contact the dean of women's office.

2. FRESHMAN CLOSING HOURS
Closing and opening hours for freshman women shall be the same as for all women students except during Rush week and Orientation week. The closing hours for these periods will be announced at the time of the students' arrival.

3. OPENING HOURS
Women may not leave their living groups before 6 a.m.

4. SIGNING OUT
Oregon State University is expected to assume full responsibility for its women students. Therefore, the university has included signing out when absent from a living group in its regulations. This practice enables university authorities to contact absent coeds in emergencies.

a. Daily Sign-outs
(1) Women must sign out at the desk when they leave if they are to be out of their living groups after 7:30 p.m. They must give complete destination address, with name of escort. (Upperclasswomen will sign out on colored cards for 11:00 p.m. closing).
(2) On Saturdays and Sundays, women must sign out if they are to be away after 2 p.m.
(3) All upperclasswomen must be signed out and must be out of the living group before 10:30 p.m. when having extended closing hours. (11:00 p.m.)

91

"substitute parents" and responsible for their female charges. They were acting in the place of the parents, *in loco parentis.*

Male students had fewer regulations, but the colleges and universities felt a similar responsibility to provide a substitute parental discipline. Alcohol on campus was either prohibited or controlled, and freshman males generally had to live in the dorms under the eye of floor counselors, who often had some association with the dean of men. Rowdy behavior would result in a call from the dean's office, and even actions off campus were subject to scrutiny and censor. For both men and women, then, the school had more than just an educational function; it was an extension of the socializing activities of the family and a continuation of adult supervision.

The sixties saw administrators, often deans of women, in conflict with students over the *in loco parentis* policies. Deans saw themselves as guardians of the mores and behaviors of their students and felt responsible as stand-in parents (even if, ironically, most of these deans were single women). The revolt against having stand-in parents dictate behavior was as violent as familial conflicts that pitted adolescents against fathers and mothers. Along with dorm hours, dress codes, and drinking regulations, choice of partner became another element in the students' rejection of the role of the institution in controlling and monitoring their activities. By the end of the decade, the *in loco parentis* philosophy had fallen into disrepute, and along with it, the injunction against interracial dating.

■ Oregon State University Student Handbook "Rook Guide," 1962–1963, pp. 90–91.

reducing her to tears by implied threats of expulsion and suggestions that girls whose morals were so debased did not belong in the institution. Repentant females occasionally found themselves among the elect, brought into the fold of the "twelfth-floor girls," bringing with them the zeal of the convert. Like many universities in the early sixties, female and male students were encouraged to embrace the freedom of the American Dream, but to keep their activities within socially prescribed boundaries.

En route to becoming a twelfth-floor girl, one coed met a fellow student who used a wheelchair. In response to some offhand remark that the disabled student made about an interest in theater but lack of opportunity to participate, the coed suggested that she could take part in a number of ways, and ultimately the girl ran lights. The girl was delighted, and everyone in the theater liked her. She, however, had never broken the curfew, as many of the theater students did, so she had never been called into Miss Skewis's office and given the treatment. She thus missed out on one of the ways of becoming a twelfth-floor girl. More significant, however, the girl had broken a much more serious rule, eliminating her from consideration. She was the girl with the blind "colored" friend.

International students may have represented a relatively small fraction of the student body, but they helped create a more cosmopolitan atmosphere than most of the Pitt undergraduates had experienced in their hometowns. The alleged removal of (at least) two biracial couples by the dean of women and her furies seemed all the more shocking and embarrassing, and it burned itself into the consciousness of the students at Pitt. Such actions appeared to be emblematic of the hypocrisy of American culture in the 1960s. What did *liberty* and *freedom* mean in such a context? The discrimination against mixed-race couples combined with similar injustices and inconsistencies, which students heard or read about, drove them into those political movements we associate with the decade.

The long hair, the beads, the peace symbols, the open display of taboo subjects, and the experimentation with sex and drugs reflected many things. Students in the sixties were, after all, a largely pampered generation. Their childhood in the fifties had allowed them to escape the grim realities of their parents' generation. Although they heard endless stories about the Depression and the war, these were just stories, like those of their immigrant grandparents, about times and situations far removed from students' daily lives. Some had fathers who may have had only two dollars in their pockets when they got married, but they now lived in suburban ranch houses, drove Buicks, and provided their children a weekly allowance. If these students found themselves short of cash, they had probably spent too much money on the weekend or on clothes. But they still had a cafeteria meal ticket and a warm bed. These students had been raised with powerful images of what America meant. Their grandparents told them stories of what it was like to live in a country that did not have the social, religious, and economic freedoms they took for granted. And that was just it. They did take it for granted. That they were free. That they all had choice. That *everyone* in *America* was free.

Clearly not, they discovered. What we now label "the sixties" was a reaction, in part, to the disillusionment many young people experienced and their associated rebellion, an attempt to set things "right."

One Student's Personal Acquaintance with "Mixing"

Let's consider in greater depth why a local Pitt undergraduate might have reacted so strongly to the rumors about the dean of women's actions. While growing up as a white male, he did not fully realize the benefits of his social position (a common tale among U.S. white males). Like many undergraduates at Pitt, he had spent his childhood in southwestern Pennsylvania, an economically depressed region that was rich in ethnic diversity but poor in tolerance. The multiethnic layers of society had stratified into a recognized hierarchy, with Old World tensions continuing in new forms. Serbs and Croats disliked one another, went to different churches, and bristled at the suggestion that they spoke the same language. (It came as no surprise to that generation when Yugoslavia fell apart in 1991.) White Anglo-Saxon Protestants (WASPs) composed, undisputedly, the top tier (with Episcopalians above the Methodists, Presbyterians above the Baptists, etc.). People of southern or eastern European origin ranked lower on the social scale—Polish and Italian Catholics were excluded, for example, from membership in the local country club (that arbiter of class status). Middle Easterners, largely Lebanese or Syrian, were lower than Italians and Poles. Jews were at the bottom. The small black population was in the sub-basement, virtually out of sight.

But it was a contested hierarchy and a complex one that took into account class as well as ethnic origin. Jews, for example, although excluded from most country clubs (and often, therefore, the major local golf course) and considered largely unacceptable for marriage by the other groups, maintained something of a middle status because of their economic position, mostly professionals, businessmen, or merchants.[2] They were admitted to the Kiwanis

2. See Annie Dillard, *An American Childhood* (New York: Harper and Row, 1987). In discussing her memory of country club subscription dances, she notes: "I wondered which of those remote country-club powers, those white-haired sincere men, those golden-haired, long-toothed,

and the Rotary. Lebanese and Syrian represented not geographical or ethnic designations but economic ones. If you had money, and you or your family came from the area now comprising Lebanon, Syria, Jordan, Iran, and Iraq, you were likely to be thought of as Lebanese. If you were of modest income and your family came from any of these places, Syrian. Irish Catholics considered themselves equal to WASPs, especially the Scotch-Irish, although the Protestant establishment did not recognize such distinctions among Italian, Irish, and Polish Catholics. Irish Catholics who wanted to join the country club had an uphill battle. You had to have a lot of money if you ate fish on Friday and wanted to play golf on the weekend.

In a small Pennsylvanian town, a "mixed couple" might refer to a Catholic/Protestant pair, but mostly it meant a Jewish/Gentile match. Interfaith marriages still remained the exception in the late fifties and early sixties. One former Pitt undergrad (Jewish) recalls that the first time he heard the term *mixed couple* was in 1956. He had arrived at the local Jewish Community Center swimming pool after lunch, as he did most days during those lazy summers of adolescence, *and there they were*: the 40ish divorced sister of a center member and her out-of-town guest, a similarly old, bearded Gentile man horsing around in the pool (40 was old to someone 12). OK, they were not making out, but part of the time they had their bodies closely pressed together. It is hard to sort out what was most shocking: the goatee (a Beatnik in the pool!); the public display of physical attraction in a desexed (more accurate, sexually repressed) family-oriented swimming pool; a Gentile in the pool (Was he circumcised? Would he change in the locker room?); or a divorced woman, old enough to be the 12-year-old's mother, acting like a teenager with hormones raging in her blood. The couple supplied a subject of conversation for weeks.

Only later did color enter into this young man's understanding of the term *mixed couple*. A tall, elegant, and intelligent black girl in his tenth-grade class and he were friends. On occasion they walked home from school together, about halfway, before their paths diverged. She was clever and funny, and they always had a good time talking. Most of the route home consisted of rural two-lane roads. As they walked, they'd sometimes get strange looks. Their walks were innocent, but unlike other Gentile girls in school, he sensed that she was *really* off-limits.

Why? Although the black population attended the fully integrated schools,

---

ironic women, had met . . . had pored over what unthinkable list of schoolchildren to discuss which schoolchildren should be asked to these dances they held for what reason. If you were part Jewish, would they find you out, like Hitler? How small a part could they detect?" (187).

interracial dating remained only a theoretical possibility in that small Pennsylvania town. He had seen mixed-race couples in New York City, where much of his extended family still lived and which he visited periodically. His grandparents belonged to the wave of eastern Europeans who had flooded the lower East Side at the turn of the century, and their descendants had spread out through the city: the upper West Side, the Bronx, Queens, and Brooklyn. On visits to the city, his parents allowed him to roam about at will, and the Village became one of his favorite haunts. In Greenwich Village, you could see *anything*: transvestites, gays, beatniks, and even biracial couples.

So that was his background. He did not consider biracial couples alien creatures, but they appeared exotic and existed on the margins of society. He did not believe that because society marginalized a group, it necessarily should be that way, or that it had to remain so. Jews, after all, were in a decidedly disadvantaged social state in western Pennsylvania. They were not as badly treated as in his parents' youth, and they were treated dramatically better than his grandparents' generation—how bad must things have been to have picked up roots, left the only home you had ever known, illegally crossed borders, and sailed across the ocean to a land where people spoke a different language and had a different culture? Jews had improved their lives, and other groups might do so as well. A natural sympathy existed in families like his for the underdog and for those on the fringes of social respectability.

The idea of a mixed couple, then, had multiple associations for him. *Mixed* mostly meant Jewish/Gentile, and that mix was one with which he was familiar. Social boundaries discouraged such mixing, but it was a fact of life. Catholic kids were taught that if they married outside the faith, their children ran the risk of forfeiting the hope of heaven, and Jewish kids were taught that an interfaith marriage was the first step to assimilation and ultimate destruction of God's Chosen People. The social chemistry of public school, however, threw teens together, and human biology often trumped the "wisdom" of their elders. (Various other ethnic and class mixes existed: Italian/Polish, Northern Italian/Southern Italian, country club/blue collar, etc., but these were less freighted with community disapproval and more an internal matter for families.) Mixed-race couples were much further removed on the social spectrum although still a part of the continuum.

At Pitt, the dorm rooms contained all manner of races and ethnic mixes, and for many students, boarding on campus marked the first time they had lived so intimately with someone of a different race (different genders and sexual orientations would come only decades later). Those students who (randomly) found themselves with a dorm partner of a different race inevitably faced questions of race and race mixing. One such pair consisted of a black student from Norfolk, Virginia, and the local white Jewish boy already introduced in this chapter. To the Virginian, his roommate was the "squarest" guy he had ever met, and the roommate, analogously, had no idea of what to make of his bunkmate. The southerner introduced his roommate to the music of Ray Charles and taught him to do the twist (the northerner was pathetic). The white boy shared with his roommate his enthusiasm for pastrami sandwiches from Weinstein's Jewish Delicatessen and Dave Brubeck's jazz. The black student was majoring in aeronautical engineering and had a slide rule on his desk; his roommate, although premed, was captivated by existential philosophy and had Camus on his. So they inhabited largely separate cultural spheres. Many college roommates experienced life with someone whose background was different. But race was an added difference for these two. The Jew thought his religious/ethnic background made him a natural ally to the black, but that was less clear to the Southerner. Race was a difficult minefield to maneuver.

The first week of the fall term, while the young black man was unpacking his things in the dorm, his roommate noticed a large bowling ball. He was surprised, for although bowling had been popular in high school, he had no expectation that it would continue and be part of his college life. He wasn't even sure there were any bowling alleys in Oakland, the university section of Pittsburgh. But for the black student, the bowling ball was symbolic. Norfolk had segregated bowling alleys, and few, if any, where he lived. Moving North for him meant moving where he could just go out and bowl wherever he wanted. The symbolism struck his roommate, who respected it.

One evening the black student returned to their room upset. He had been bowling and had been harassed by some rednecks. In reaction, he said that he preferred living in the South because at least there the racial lines were clearly drawn and everyone understood them, whereas in the North, there were no clear lines, and he kept bumping into walls. It was the hypocrisy of the situation that so got to him. His disillusionment with the North hurt and shocked his white roommate, who felt shamed that the "liberal" North had let his roommate down. And, in thinking about the reality of that life, of living as a black in the North, he came to think about and consider the scarcity of mixed couples. It was a topic that he discussed with his family during Thanksgiving break his first year at Pitt.

His father loved politics, and it was one of his favorite topics of conversation at the dinner table. Although he considered himself progressive, at times caution overcame sentiment. He had warned his son about which political groups not to join, a reflection of the hard lessons he had learned during the

McCarthy era. They argued, but the son understood where he was coming from. What he was not prepared for was his father's response to a hypothetical question about interracial dating and marriage. His father's reaction stunned him. A couple years later, the Pitt student would see Spencer Tracy in *Guess Who's Coming to Dinner?* Tracy looked rather like his father and practically repeated the judgment he had heard as a freshman: "bad idea." The couple would not be accepted by either "community," and what would happen to the children? Wasn't life hard enough without adding extra burdens? There was no blatant racism, but a pragmatic argument based on the social realities of the time. The son, of course, argued in the manner of a student of the sixties: how would society change if it remained locked in the status quo? Would Jews be allowed in medical schools today if there hadn't been that pushy first generation? It was an argument that his generation had with its parents and would continue to have in different forms for many years. He and his peers had been raised to believe in acceptance and openness. They had not lived through the economically stressed thirties or the war years. They did not remember the faces of those lost to the Holocaust. They knew little, or nothing, of lynching in the South, or the race riot in Tulsa. They did know the American Dream. Their schools had emphasized it throughout the fifties, and they had accepted that vision: America, the land of the free and of unlimited opportunity for all.

This student's experience illustrates the sensitivity of the subject of race, and that mixed-race couples were scarce because of prejudicial biases that had deep cultural roots reinforced by social and economic forces, even in the most progressive regions of the country. The resulting discrimination worked to maintain a status quo that privileged the whites and exploited the rest. In Pennsylvania a strong labor movement had improved the condition of the working man. But in the early sixties, gender and class remained strong social markers that determined status, wealth, and power. Race was even more deeply ingrained. The black population had not shared the upward mobility of Slavs, southern Europeans, and Jews after the war. August Wilson's plays reflect the frustration that the black community felt during those years, and although Pittsburgh did not explode like some other northern cities, a noticeable tension was present. Places like the Crawford Grill jazz spot continued to provide a space where whites and blacks could listen to music together. However, throughout the sixties, racial strife mounted, and street life reflected it. By the summer of '65, white boys did not park more than a block away from the jazz club.

In Class

Progressive forces were at work. At the University of Pittsburgh, numerous courses reinforced the melting pot mythology that informed parts of the high school social science curriculum at the time as well as the relatively liberal values of many students. In anthropology classes, students learned about the relativity of cultural values, and about the "myth" of racial difference. In zoology, they discovered that all human groups were interfertile (i.e., in a scientific sense, they belonged to the same species), and that popular racial categories carried little biological meaning. Their textbooks stated that race mixing had been a constant process in the evolution of modern societies and had been a consistent feature in ancient history as well. Cultural anthropology (Anthropology 80), a large (300 students) general education course taught by the popular lecturer Arthur Tuden, dealt extensively with ethnicity and race.[3] In his first lecture of the term, Tuden contrasted anthropology with the other social sciences and explained that whereas subjects like sociology and economics tended to have a Western orientation, anthropology used non-Western sources most of the time. He stressed that from the study of these cultures, we learn that everyone thinks his culture is the best. An implication of this recognition is that we must judge values in their context, not across cultures. It is impossible to be objective cross-culturally because we all unconsciously carry our own cultural values and judge others according to them.

This idea of cultural relativity resonated with Tuden's students for several reasons. Many had grown up in ethnically diverse neighborhoods, each group with its own conviction of its superiority and condescension for the rest, and the concept of cultural relativism provided the students a perspective that acknowledged the legitimate status of ethnic pride while undercutting any legitimatization of discrimination. The notion that different cultures had different accepted practices, and that what might be shocking to Americans could be normal to others, also appealed to their liberal sense of respect for other cultures. Tuden used the examples of female circumcision or eating maggots. These might cause cultural shock in those outside the culture, but students were told they needed to recognize their emotional reactions as normal cross-cultural expressions of ethnocentrism.

As a background to the cultural anthropology that constituted the majority of the course, Tuden spent several weeks on physical anthropology, beginning with a discussion of the evolution of humans. Students were given a handout

3. Class notes, Farber Papers, 1963.

on the relative chronology of Paleolithic man taken from Ashley Montagu's *Introduction to Physical Anthropology* (3rd edition, 1960). The final portion of the section on physical anthropology dealt with race, the physical variation in modern humans during the past 35,000 years. Anthropologists recognized race, according to Tuden, by outward appearance, not genetic difference. In that sense, the commonly described races consist of artificial categories and have no biological meaning. In theory, one could divide humans by looking at groups that vary in their genetic makeup, and here he mentioned the famous geneticist Theodosius Dobzhansky, who distinguished between cultural and environmental differences on the one hand and hereditary differences on the other. What about the comparative relationships of races, and about race mixing? Tuden was clear that no evidence existed for any racial superiority in intelligence among races; any observed or alleged differences were due to cultural and environmental factors. The New World had experienced extensive race mixing in the past 400 years (Indian, Negro, Caucasian). In Latin America, he claimed, there was no racial consciousness; differences were based on culture, whereas in the United States, anyone who had one-eighth or more of Negro ancestry was considered to be of a different race than Caucasians. He predicted, however, that in 300 years, there would be no more races due to the continued mixing that was occurring. And, of course, these basic groupings were artificial to begin with, so the "mixing" was more of a social phenomenon than a biological one.

Assigned readings reinforced points made in lecture. Alfred Romer's *Man and the Vertebrates* (first published in 1933) clearly stated that human races mix and have mixed in the past, and that "pure" races are manmade ideals. The more up-to-date (1958) textbook, A. E. Hoebel's *Man in the Primitive World*, gave additional genetic information and provided a biological perspective on race. The author mentioned that some anthropologists consider race a myth but recognized that the concept had been around for a long time and was likely to continue. Students, however, were cautioned to be careful in how they conceptualized race. Variation of most traits—for example, height—was greater within a race than between races. The commonly recognized races were based solely on external appearances with little or no knowledge of their genetic basis. Leading anthropologists like Sherwood Washburn advocated using the theory of evolution to clarify the relationships of various human groups. The application of genetic principles to the study promised to make it possible to create a deeper and more meaningful understanding of race. But, that was still something for the future. At present, Tuden cautioned his students that they needed to employ terms like race with a grain of salt.

Scientism

If you had been a young boy or girl in the middle decades of the twentieth century who watched TV on Saturday mornings, you would likely have seen a program called *Watch Mr. Wizard*. It was the brain child of Don Herbert, who ran the show starting in 1951 and entertained his audience with an engaging set of gee whiz experiments, often with materials found around the home, intended to demonstrate various principles about the natural world. The show was a big hit, and the term Mr. Wizard became part of the vocabulary of the decade. It also reflected the pervasively positive attitude toward science. Science and technology promised to create a better and more prosperous life for what would become the baby boom generation. When the Soviets launched their space satellite Sputnik in 1957, and the U.S. government

wanted to accelerate the growth of science and technology in American schools, it was able to draw on this fascination with science that had become so deeply ingrained. The confidence in science's ability to solve problems was an integral component of fifties and sixties thinking, and historians often talk about the *scientism* of the period.

Scientism refers to the view that science provides the best methods for answering questions about the natural world, social problems, and cultural issues. Scientism, of course, did not begin in the 1950s; it has had a long history, going back at least to the eighteenth century, and is often associated with various schools of positivist philosophy as well as with the broader Enlightenment agenda. In the United States, forms of scientism became quite popular during mid-century, when "scientific" methods, and technological fixes, seemed to promise solutions to important

problems facing Americans in the postwar era: in housing, agriculture, transportation, industry, advertising, medicine, child psychology, and education. In retrospect, our overconfidence in science to solve so many perplexing issues looks a bit naïve today, and the very term *scientism* has taken on a somewhat pejorative connotation. The various revolutions of the 1960s, with their questioning of all authority and rationality, and the emergence, also beginning in the late sixties, of various post-positivist philosophies, have undercut the often simplistic ways that problems were framed in the fifties. It is difficult to imagine anyone today not realizing that environmental challenges have social as well as scientific dimensions, or that politics play a role in biomedical policy.

Despite the onslaught of postmodernism and widespread public skepticism today about many of the universally accepted ideas in science—let's not forget that polls suggest half the population in this country doesn't accept the theory of evolution—we still find strong strains of scientism in the United States. Evolutionary psychologists, sociobiologists, and a number of philosophers of science argue that biology can provide a foundation for ethics and can be used to answer ethical problems that emerge in biomedical research and technology, as well as in more traditional areas of the life sciences.

■ *Left,* Don Herbert (Mr. Wizard). Courtesy of Mr. Wizard Studios, Inc. *Right,* Sputnik 1. Replica of Sputnik 1 on display at the United Nations. National Air and Space Museum, Smithsonian Institution (SI 77-1022).

Another popular class that touched on the subject of race mixing was general genetics (Genetics 35).[4] The sixties was an exciting time to be studying this rapidly expanding subject. Elliot Spiess, the professor, had been a student of Theodosius Dobzhansky, and one of the books on his reading list for the class was Dunn and Dobzhansky's *Heredity, Race, and Society*, as well as Dobzhansky's *Mankind Evolving*. The course focused mostly on the genetics of the fruit fly (*Drosophila*), which had been the model organism for much of the work done in the previous couple of decades. Human genetics occupied the last two weeks and functioned to introduce the principles of population genetics. The cornerstone in understanding populations was the Hardy-Weinberg law, and Spiess made it the centerpiece of his discussion. Although it is a simple algebraic formula, it leads to counterintuitive conclusions, the most significant being the surprising stability of genetic equilibrium. Why don't the dominant genes in a population ultimately swamp the recessives? If the gene for red flowers is dominant over the gene for white flowers, and if, when we cross a red-flowered plant with a white-flowered plant, offspring are all red-flowered plants (and in the next generation of "mixed" plants, the ratio of red to white is 3:1), then wouldn't we expect that in time we would only very, very rarely run across a white flower? But no, the ratio of red to white in mixed crosses remains constant: 3 to 1. Generation after generation.

What about attempts to improve populations? Could we, through selection, eliminate genes for undesirable traits, say, ones that would result in an individual who could not survive until adulthood? If they are recessive (like the white color in the flower mentioned above), the Hardy-Weinberg law suggests that a residual pool of these genes is likely to continue circulating. Consequently, students were informed that earlier visions of transforming the human population into one that was free of deleterious genes was doomed to failure. (Genetic counseling might allow *individuals* who had a risk of carrying a particular trait to think twice about having offspring, but it could never eliminate the genes completely from the *population*.)

The lectures and readings in courses like anthropology and genetics held important implications. Undergraduates were being taught that our everyday concepts of race were more a social construction than a discrete biological reality. These constructed human races inherently mixed, and no reason existed to assume one could ever arrange races into a hierarchy, or to conclude that there was any valid biological reason to worry about race mixing. The very term *race mixing* seemed to be a concept that had more social than biological meaning.

4. Class notes, Farber Papers, 1965.

The classes led students with a science orientation to wonder about the many uses to which these terms were applied, and more important, to worry about how carelessly people bandied them about. Teachers, like Tuden or Spiess, often contrasted popular myths with more objective knowledge and enjoined their students to be critical in their thinking and evaluations. Anthropology stressed the arbitrary nature of many social conventions. And, genetics made them realize how humans were all part of a large gene pool consisting of many subpopulations that continually exchanged genes, so that the entire pool was in a dynamic state. Their teachers stressed the importance of applying this knowledge correctly to the world.

Even for students without a scientific orientation, the lessons from such undergraduate general education courses carried important social significance. During the early sixties, science held considerable credibility. In part, this was a continuation of the scientism that had characterized much of the fifties. The launching in October 1957 of the Russian space satellite Sputnik sparked a national drive to improve science education in the United States. Americans were shocked that the Soviet Union had beaten them in the first stage of the space race, and many concluded that the country had to concentrate on science and mathematics if the United States wished to maintain the technological superiority it had held since the end of the Second World War, and, more important, if the country did not wish to lose the current Cold War. The race to space was symbolic of technological sophistication, but it equally suggested strategic advantage: a rocket powerful enough to escape Earth's atmosphere could be powerful enough to deliver an atomic bomb to New York, Pittsburgh, or Los Angeles. The Russians followed Sputnik 1 with Sputnik 2 in November (partly in celebration of the 40th anniversary of the Russian Revolution). This larger satellite even carried a dog (Laika) as passenger, suggesting to many that man could survive in space. These impressive feats gave U.S. officials anxiety, made all the worse by the failure of America's first attempt to launch a satellite in December of the same year. A Vanguard rocket exploded after briefly leaving its launch pad in Cape Canaveral (with the full press corps watching). Congress quickly passed the National Defense Education Act of 1958, which included more than a billion dollars to improve science education, and with funds and attention, science became increasingly a subject that held status and respect. What scientists said was taken seriously and not discounted easily.

Undergraduates listened to lectures on anthropology and genetics in this environment. The subjects were sufficiently important that the United States invested defense dollars in them. When we read about the sixties today, we often see mention of the "counterculture" and the rejection of science in favor of more spiritual (often drug-induced) insights. But that reflected the late six-ties, and was partly a reaction to the enthusiasm and support of science earlier in the decade. During the first half of the decade, the new knowledge in the life sciences, particularly genetics, seemed reliable and important, holding impli-cations for society. We take a lot of it for granted today, but at the time, people regarded it as novel. This was all the more true since earlier in the century, a quite different set of scientific ideas had dominated. The new genetics and physical anthropology contradicted many ideas about race and race mixing, and it is worth considering in what ways the new approaches informed the six-ties generation.

2 Scientific Ideas on Race Mixing

A young couple in the early 1960s would not likely have consulted a biology book to help decide whether their different racial backgrounds posed an obstacle to getting married and raising a mixed-race family. Even so, what passed for scientific opinion could easily have played a part in their decision, and in the reactions of their families. In many subtle and some not so subtle ways, scientific judgments influenced individual choices, social acceptance, and legal constraints. At the time, seventeen states had anti-miscegenation laws that prohibited marriages involving people of certain different races, and an extensive body of literature justified those laws by reference to science.[1] Even so, in the three decades before *Loving v. Virginia*, a shift in thinking about interracial marriage occurred in the United States. In part, that shift reflected a new social landscape altered by World War II and the civil rights movement. Old discriminatory attitudes and biases met resistance from a new generation that rejected the racism that had been so widely accepted and assumed in the early decades of the century. Students on campuses would have been aware of the altered intellectual environment in which discussions of race mixing took place. They were also influenced by a new scientific perspective on race mixing to which they had been exposed in the classroom. Science had a lot to say about race, and many professors made efforts to elaborate on what they

---

1. In 1967, but before the Supreme Court decision of *Loving v. Virginia* was handed down, one of the seventeen states (Maryland) with existing anti-miscegenation laws repealed its law. Fourteen other states had repealed similar laws in the fifteen years before the decision. For details, see Werner Sollors, ed., *Interracialism: Black-White Intermarriage in American History, Literature, and the Law* (Oxford, Oxford University Press, 2000), p. 30. The word *miscegenation* comes from a political hoax. It was the title of an anonymous pamphlet that coined the new word (from Latin, "to mix" and "race") and argued for its virtues. A New York journalist (David Croly) wrote the pamphlet and printed it in order to inject the issue of race mixing into the political campaign of 1864. His aim was to embarrass the Republican Party by making it appear that they supported race mixing. See Sidney Kaplan, "The Miscegenation Issue in the Election of 1864," *Journal of Negro History* 34, no. 3 (1949): 274–343.

saw as the social implications of contemporary knowledge in the life sciences, especially the transformation of attitudes on race mixing.[2]

"Race"

Terms in science often have a technical meaning that can differ considerably from everyday uses of the word. To make matters even more complicated, the meanings of both scientific terms and their everyday usages can change over time. It is important, then, before discussing the changing ideas and attitudes on race mixing to consider briefly what race has meant in scientific literature.

The term *race* refers to one of several basic concepts in classification. Naturalists classify plants and animals for many reasons. Primarily, however, they do so to organize information (how would you find the properties of one of a half million insects without some method of information retrieval?). The organization of the material can be strictly for convenience, or, more ambitiously, it can aspire to reflect some perceived order in nature. Classifications, therefore, can vary greatly, and to understand a particular classification one needs to inquire about the purpose for which it is designed. In the Renaissance, authors organized animals alphabetically, and it worked just fine because the number of animals known to natural history remained small (a few hundred). In later centuries, when the numbers of animals (and especially plants) increased, alphabetical classifications appeared completely impractical, and naturalists developed other methods.

Beginning in the second half of the eighteenth century, and increasingly in the early nineteenth century, the term *race* became a standard concept in the classification of animals. Naturalists were coming into possession of vast quantities of data because of increased exploration, colonization, and the consequent expansion of museum collections. Work done in the field added to the information available in museums and private collections. This expanded information base led to an appreciation of the range of variability within species. Large collections had multiple specimens of the same species, and naturalists noted that these specimens varied in interesting ways: color and size, what time of year they had been obtained (i.e., traits based on seasonal changes), age (i.e., exhibiting life stage differences), sex, and geographical origin.

2. William Provine has argued that scientists altered their opinion between the 1930s and fifties, but that they did so for social and political reasons rather than for any scientific ones. See William Provine, "Geneticists and the Biology of Race Crossing," *Science* 182 (1973): 790–96. Although miscegenation referred to mixing of different races, this study will consider only White/Black race mixing, which constitutes the largest percentage of race mixing and has generated the most attention.

Naturalists like Charles Darwin's correspondent in India, Edward Blyth, for example, devoted considerable time and effort to working out definitions to encompass the patterns of difference found in nature, and to distinguish minor variation from more significant varieties. He titled one of his most famous articles (published in 1835), "An Attempt to Classify the 'Varieties' of Animals, with Observations on the Marked Seasonal and Other Changes which Naturally Take Place in Various British Species, and which Do Not Constitute Varieties."

Naturalists, like Blyth, focused particularly on varieties, or races, of organisms. These generally referred to groups of organisms that possessed some physical characteristics that distinguished them from the more common type. The varieties usually were separated geographically from other members of the species, and therefore were often referred to as *geographical races.* Naturalists had difficulty in determining if the varieties constituted different species or merely represented a subspecies (or a smaller subdivision). Considerable confusion resulted when some naturalists identified a group as a separate (or new) species while others lumped it as a variety with an already known species. Additional confusion was added by naturalists not using the terms *variety, race,* and *subspecies* consistently. Virtually all those engaged in classification (and that would be a majority of the naturalists) had to contend with these routine problems in the decades straddling the middle of the nineteenth century.

Darwin grappled with this issue in the 1850s while he was working on the classification of barnacles and, more important, while elaborating his new theory. Understanding varieties, and their origins, was the key to unlocking the "mystery of mysteries": the origin of species. The full title of his classic book, reflected the importance of the subject: *On the Origin of Species by Means of Natural Selection, or the Preservation of Favoured Races in the Struggle for Life.* How did races come into being? What was their relationship to other members of the same species? And most important, in what ways were they incipient new species? Those questions lay at the heart of Darwin's theory.

So, the concept of race has been central in the history of the life sciences. What about human races? Since roughly the middle of the eighteenth century, naturalists have described races of humans. The famous French Enlightenment naturalist Georges Louis Leclerc, comte de Buffon, first used the term *race* in a scientific sense. In the opening three volumes of his massive *Histoire naturelle* (1749), he published essays that discussed the philosophical foundations of classification and treated humans as subjects of natural history. Buffon argued against other naturalists who wished to define species in terms of essences, that is, by a set of essential characteristics (a tradition going back to

Aristotle). Instead, he claimed that naturalists used such categories for convenience. This stance reflected his attraction to the "new science" based on
Newton and Locke, popular with his fellow *philosophes*. Buffon stressed the
diversity of form, and when he described individual species, as he did in the
hundreds of articles within his natural history, he made a point of detailing
the various geographical varieties that were known. For Buffon, the environment played a central role in determining the appearance of animals, and in
humans, he recognized six "races" whose appearance reflected their geographical location. These races, like other animal varieties, he claimed, were fully
interfertile and somewhat plastic. That is, their physical characteristics did
not represent a fixed set of traits. A change in the environment or a geographical shift could result in physical alterations. He believed that various human
groups had "degenerated" from a pristine first type (suspiciously French looking). Buffon also recognized that his divisions were somewhat arbitrary and
used for convenience, as were many such concepts in classification. That is,
one could divide humans into 5, 9, 14, or 36 groups if one chose. Drawing on
Buffon, Johann Friedrich Blumenbach in 1775 published the first edition of his
De Generis Humani Varietate Nativa, a work that became the standard beginning reference point for discussions of human races. Like Buffon, Blumenbach
stated that he constructed his classification as a convenience, and that dividing
humans into races did not undercut the basic unity of the human species.

The notion that racial categories were somewhat arbitrary reflected the recognition by naturalists of the great amount of variation among human groups.
Within Europe, those who lived in the northern countries were thought to
look quite different from those in the southern regions of the continent. The
distinctions were based mostly on skin color, hair type, eye color, skull shape,
and overall body shape; however, temperament and intelligence were also considered.

In the generation after Buffon and Blumenbach, most naturalists came to
believe that through comparing the basic anatomy of different animals, they
could distinguish different species on basic morphological characteristics. The
most influential naturalist in this movement was Georges Cuvier, the famous
comparative anatomist at the Paris national museum. Cuvier claimed not only
that could he rigorously define species, but that they were fixed—they could
not change in time. Races were merely geographical varieties. It was Cuvier's
conception of the fixed nature of species that Darwin overturned in his famous
theory of evolution. For Darwin, a race, or geographical variety, came into
being through a process he called *natural selection*. These varieties had the
potential to become new species if they developed greater and greater differ-

ences. The process, however, was complex, and consequently, distinguishing between a variety and a new species could be quite difficult. The ability to reproduce successfully was a rough guide: hybrid infertility or reduced fertility appeared gradual in varieties over a large range, and experienced naturalists could draw tentative lines based on this information.

Evolutionary change came about *very* slowly. Naturalists in Darwin's day, and in the generation that followed him, felt justified, therefore, to continue much of what they had been doing in classification—that is, collecting specimens, carefully observing and measuring them, and preparing detailed anatomical descriptions of different species (and varieties or races). Evolution explained why species looked as they did, why so much diversity existed on the planet, and why large groups of animals shared many basic characteristics. Evolution also opened up the possibility of understanding how life had developed on Earth and gave new insight into animal relationships. Nonetheless, a lot of the day-to-day collecting, describing, and arranging continued in a manner that had been going on for decades.

What about human races? Here the story gets complicated because a lot more was at stake. In discussing animals, naturalists had no sense of a hierarchy among races—that is, one geographical race was as "good" as any other: what distinguished them was that they were adapted to specific regions, and in that sense, each race or variety was best suited for its own natural habitat. But when it came to humans, cultural factors intruded, and European naturalists, later anthropologists, did not hesitate to comment on the "superior" or "inferior" traits that characterized different human races. This was hardly surprising. The English had long held the Irish as inferior and regarded the dark-skinned people of Africa even more so. Whether slavery was justified by the racism of Europeans or the consolidation of a system of slavery led to increased racism has been a subject of debate among historians. Whatever the cause, Europeans regarded the natives of the New World and the inhabitants of Africa as inferior to themselves. Cuvier, who believed that humans could be divided into three main races, held that the Caucasian race was aesthetically and intellectually superior to the other two (Mongolian and Ethiopian).

Although some naturalists wanted to argue that the major races of humans constituted different species—had originated separately—the general belief that humans were of one species prevailed. The alternative view remained popular with supporters of slavery, but despite some famous naturalists who supported the view—Louis Agassiz, for example—strong religious and scientific traditions continued to buttress the idea of a single origin of man. But not necessarily the idea that different races were equal. Perhaps the best example of how this manifested itself can be found in the pervasive belief that race mixing was unnatural and unwise. Many writers stated that the result of such mixture, particularly mulattoes resulting from Caucasian and Negro crosses, were doomed to be degenerate.

Early Twentieth-Century Ideas on Race Mixing

Attitudes about race mixing in the early part of the twentieth century reflected broader notions on race and particularly the "scientific racism" that had come to dominate discussions of race.[3] Naturalists, and later physical anthropologists, used the techniques of zoology to define the physical characteristics of humans groups living in different parts of the globe. Little agreement existed about how best to divide humankind, or into how many groups or subgroups. But educated opinion held a general sense that the divisions of Blumenbach

3. An extensive literature on the history of scientific racism exists. As background to this study, see especially Elazar Barkan, *The Retreat of Scientific Racism: Changing Concepts of Race in Britain and the United States between the World Wars* (Cambridge: Cambridge University Press, 1992); Mark Haller, *Eugenics: Hereditarian Attitudes in American Thought* (New Brunswick, N.J.: Rutgers University Press, 1984); Nancy Stepan, *The Idea of Race in Science: Great Britain 1800–1960* (Hamden, Conn.: Archon Books, 1982); Pat Shipman, *The Evolution of Racism: Human Differences and the Use and Abuse of Science* (New York: Simon and Schuster, 1994); Audrey Smedley, *Race in North America: Origin and Evolution of a Worldview* (Boulder, Colo.: Westview Press, 1993); and Sandra Harding, ed., *The "Racial" Economy of Science: Toward a Democratic Future* (Bloomington: Indiana University Press, 1993). Gunnar Mrydal, *An American Dilemma: The Negro Problem and Modern Democracy* (New York: Harper and Row, 1944), states the moral inconsistency between the American creed and the treatment of Negroes with great force and influence. For an interesting discussion of the background of Myrdal's study, see David Southern, *Gunnar Myrdal and Black-White Relations* (Baton Rouge: Louisiana State University Press, 1987).

The literature on race mixing is substantially smaller than that on race or on scientific racism. Particularly useful are Rachel Moran, *Interracial Intimacy: The Regulation of Race and Romance* (Chicago: University of Chicago Press, 2001); Paul Spickard, *Mixed Blood: Intermarriage and Ethnic Identity in Twentieth-Century America* (Madison: University of Wisconsin Press, 1989); Joel Williamson, *New People: Miscegenation and Mulattoes in the United States* (New York: Free Press, 1980); Joseph Washington, Jr., *Marriage in Black and White* (Boston: Beacon Press, 1970); Peter Wallenstein, *Tell the Court I Love My Wife: Race, Marriage, and Law—An American Story* (New York: Palgrave Macmillan, 2002); Randall Kennedy, *Interracial Intimacies: Sex, Marriage, Identity, and Adoption* (New York: Pantheon, 2003); Peggy Pascoe, *What Comes Naturally: Miscegenation Law and the Making of Race in America* (New York: Oxford University Press, 2009); Carl Degler, *Neither Black nor White* (New York: Macmillan Company, 1971); Renee Romano, *Race Mixing: Black-White Marriage in Postwar America* (Cambridge: Harvard University Press, 2003); David Hollinger, "Amalgamation and Hypodescent: The Question of Ethnoracial Mixture in the History of the United States," *American Historical Review* 108, no. 5 (2003): 1363–90; Werner Sollors, *Neither Black Nor White Yet Both* (Oxford: Oxford University Press, 1997), and his edited volume, *Interracialism: Black-White Intermarriage in American History, Literature, and the Law* (Oxford: Oxford University Press, 2000).

or Cuvier provided a good guide, and that, depending on the context, finer subdivisions could be enumerated. Physical characteristics, however, were not the only defining traits of the races defined. To those, writers added cultural, behavioral, and moral ones. Naturalists regarded races as constituting a general hierarchy, and these rankings came to be of considerable importance in justifying the actions of governments, both domestically and in colonial affairs.

Life scientists in the first three decades of the twentieth century were mostly negative in their remarks on race mixing in humans. Scientists discussed the subject primarily in the context of eugenics, a movement that sought to enlighten the public about the implications of research in genetics and on human heredity, purportedly from an evolutionary perspective.[4] Such studies demonstrated, eugenicists claimed, that physical and mental traits resulted from inherited rather than environmental factors.[5] If defective individuals continued to reproduce, they argued, the future population of the country would be unfit for survival. The goal of eugenics, therefore, was to educate the public and seek legislation that would guarantee that "the fit only shall live."[6] This injunction did not mean extermination of those deemed unfit, but meant "the exclusion of those, as parents, who are incapable of creating fit children."[7] Eugenicists worked to educate Americans in prudent mate selection, informed by the latest knowledge in genetics. Of central concern to the eugenics movement in the United States was what its proponents took to be a rising tide of the "feebleminded." The movement promoted state sterilization laws that limited the

4. Many studies of the eugenics movement exist. Useful background for this book can be found in Garland Allen, "The Eugenics Record Office at Cold Spring Harbor, 1910–1940: An Essay in Institutional History," *Osiris*, 2nd ser., 2 (1986): 225–64; Diane Paul, *Controlling Human Heredity: 1865 to the Present* (Atlantic Highlands, N.J.: Humanities Press, 1995); Daniel Kevles, *In the Name of Eugenics: Genetics and the Uses of Human Heredity* (Berkeley: University of California Press, 1985); Hamilton Cravens, *The Triumph of Evolution: American Scientists and the Heredity-Evolution Controversy, 1900–1941* (Philadelphia: University of Pennsylvania Press, 1978), Haller, *Eugenics*; Edward Larson, *Sex, Race, and Science: Eugenics in the Deep South* (Baltimore: Johns Hopkins University Press, 1995); Kenneth Ludmerer, *Genetics and American Society: A Historical Appraisal* (Baltimore: John Hopkins University Press, 1972); Lyndsay Farrall, *The Origins and Growth of the English Eugenics Movement, 1865–1925* (New York: Garland Publishing, 1985); Wendy Kline, *Building a Better Race: Gender, Sexuality, and Eugenics from the Turn of the Century to the Baby Boom* (Berkeley: University of California Press, 2001); and Mark Largent, "'The Greatest Curse of the Race': Eugenic Sterilization in Oregon, 1909–1964," *Oregon Historical Quarterly* 103, no. 2 (2002): 188–209.

5. The nature/nurture debate was popularized by the eugenics movement, but it was also part of an older, and continuing, debate over Lamarckian versus Darwinian evolution.

6. Grant Hague, *The Eugenic Marriage: A Personal Guide to the New Science of Better Living and Better Babies* (New York: Review of Reviews Company, 1914), 1:10.

7. Hague, *The Eugenic Marriage*, 1:10.

fertility of those deemed unfit to reproduce and encouraged "fit" families to produce many offspring.[8] It also strenuously sought to restrict the wave of immigration from non-Anglo-Saxon countries, for eugenics reformers held that the influx of these "inferior" individuals threatened to dilute the "American race." (*Race* at the time denoted what today would be called *ethnic groups*, e.g., Irish, Jewish, Italian, Serbian, as well as what today would be termed races, e.g., Caucasian, Negro). Concern about race mixing, since it carried the specter of "race suicide," was an important dimension of the eugenics program. For the most part, however, the eugenics literature focused more on the problem of feeble-mindedness in white populations and the threat of immigration from "inferior" European countries than on race mixing of "Negroes" and "Caucasians," not because eugenicists approved of such alliances, but because social, and legal, forces already discouraged such mixing.[9]

Eugenicists discussing race mixing of Negroes and whites did so primarily in discouraging terms. Scientists perceived two main problems. The first drew on the ranking of races, a well-established tradition in European and American anthropological literature.[10] On the basis of their reading of a group's historical "accomplishments," Anglo-American writers ranked "Anglo-Saxons" among the highest, southern and eastern Europeans lower, the Irish below them, and Africans lower still. "Dilution" of the "American race" (i.e., Anglo-Saxon) through race mixing could only lower the potential of future generations, and according to a number of authors, it therefore should be avoided.[11] Anglo-

---

8. For interesting studies of the sterilization laws, see Philip Reilly, *The Surgical Solution: A History of Involuntary Sterilization in the United States* (Baltimore: John Hopkins University Press, 1991), and Mark Largent, *Breeding Contempt: The History of Coerced Sterilization in the United States* (New Brunswick, N.Y.: Rutgers University Press, 2007). One of the most popular eugenics texts, Paul Popenoe and Roswell Hill Johnson, *Applied Eugenics* (New York: Macmillan, 1920), has two chapters: "Increasing Marriages of Superiors" and "Increasing Birth-Rate of Superiors."

9. Larson, *Sex, Race, and Science*, states that the eugenics movement in the South was not especially concerned about blacks, or about miscegenation, which was illegal. This, of course, is not to suggest extensive race mixing hadn't been taking place since early times among European settlers, Indians, and Africans (slaves and free).

10. See John Haller, Jr., *Outcasts from Evolution: Scientific Attitudes of Racial Inferiority, 1859–1900* (Urbana: University of Illinois Press, 1971), and George W. Stocking, Jr., *Victorian Anthropology* (New York: Free Press, 1987), for the background of these rankings.

11. Madison Grant was among the most outspoken on the dangers of immigration. See his *The Passing of the Great Race* (New York: C. Scriber, 1916). On Grant, see Jonathan Peter Spiro, *Defending the Master Race: Conservation, Eugenics, and the Legacy of Madison Grant* (Burlington: University of Vermont Press, 2009). On immigration, also see Nathan Fasten, *Origin Through Evolution* (New York: Alfred Knopf, 1929). Although ranking of races was a popular notion, there was also a strong sense among the eugenicists that the bulk of immigrants from Europe were from

Saxon crosses with Negroes were thought to be worse than mixes with Slavs or Sicilians, since those of African descent were lower on the scale in human accomplishments. As one distinguished geneticist, Edward East, wrote, "In reality the Negro is inferior to the white. This is not hypothesis or supposition; it is a crude statement of actual fact. The Negro has given the world no original contribution of high merit. By his own initiative in his original habitat, he has never risen. Transplanted to a new environment, as in the case of Haiti, he has done no better. In competition with the white race, he has failed to approach its standard."[12]

The second main problem concerned the fear that alliances of individuals from distant groups (e.g., Anglo-Saxon and African) ran the risk of producing "disharmonious crosses." The idea had been popularized by the Norwegian Jon Mjoen, who studied the offspring of Norwegians and Lapps in northern Norway. Mjoen claimed that these hybrids showed more disharmonies than the parent stocks. Some were minor (disproportionate extremities, or large ears), but others appeared more serious (higher rates of diabetes, lower resistance to tuberculosis). He believed that the higher rates of diabetes in Lapp-Norwegian hybrids reflected "glandular lesions" resulting from hereditary disharmonies.[13] Medical writers supported the view that crosses involving individuals from the "primary races" might give rise to "chaotic constitutions." A review article in 1933 summarized: "It may be said that the bulk of medical opinion is against hybridization between the Primary Races and that the best eugenic opinion is definitely against it."[14]

The study of plant and animal breeding reinforced concerns about the biological ill effects of race mixing. Breeders had long worked to establish pure lines, which had economic value and would breed true. Edward East's widely

---

the lowest strata of the countries from which they came. This made the situation even worse. See Popenoe and Johnson, *Applied Eugenics*.

12. Edward East and Donald Jones, *Inbreeding and Outbreeding: Their Genetic and Sociological Significance* (Philadelphia: Lippincott, 1919), 253. East wrote the final chapters dealing with human race mixing, so I shall refer to East as its author in what follows.

13. See Jon Alfred Mjoen, "Harmonic and Disharmonic Race-crossings," *Scientific Papers of the Second International Congress of Eugenics* (Baltimore: Williams and Wilkins, 1923), 41–61. This meeting took place at the American Museum of Natural History in New York, and consequently Mjoen's views became widely known in the United States. Also see his "Biological Consequences of Race-Crossing," *Journal of Heredity* 17, no. 5 (1926): 175–85, and "Race-Crossing and Glands: Some Human Hybrids and Their Parent Stocks," *Eugenics Review* 23 (1931): 31–40.

14. K. B. Aikman, "Race Mixture," *Eugenics Review* 25, 3 (1933): 161. The position was contested by the geneticist William Castle, but it is not clear that he had much effect at the time. See Provine, "Geneticists and the Biology of Race Crossing," 792. Also, see Paul, *Controlling Human Heredity*, 112–13.

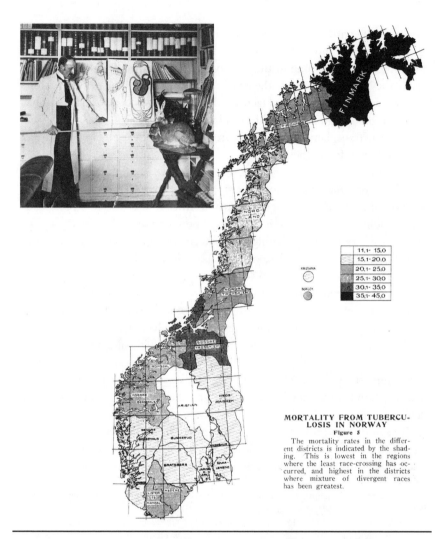

MORTALITY FROM TUBERCU-
LOSIS IN NORWAY
Figure 5

The mortality rates in the different districts is indicated by the shad-
ing. This is lowest in the regions
where the least race-crossing has occurred, and highest in the districts
where mixture of divergent races
has been greatest.

Alfred Mjoen and His Norwegian Research on Mixed Races

American eugenicists often cited Alfred Mjoen's research on Norwegian and Lapp mixed-race offspring, which claimed that higher rates of disease characterized the "hybrids." He included this map in a research article as evidence of the greater rates of tuberculosis among the offspring of Norwegian and Lapp crosses.

■ *Left*, Alfred Mjoen. Courtesy of the American Philosophical Society. *Right*, From Alfred Mjoen, "Biological Consequences of Race-Crossing," *Journal of Heredity* 17, no. 5 (1926): 179.

read *Inbreeding and Outbreeding* argued that inbreeding can be used to isolate traits found in parental stocks. If rigorous selection is applied, a breed can be developed with highly desirable traits and potentially commercial value. What about crossbreeding? According to East, it represented an even better and more widespread method of improving stocks. "Hybrid vigor," or "heterosis," he noted, had given rise to impressive gains in horticulture and animal husbandry. One might think that East's enthusiasm for "hybridization" in corn and hogs might have inclined him to consider race mixing in humans as potentially positive, and he did to some extent. East devoted two chapters of his book to the consideration of inbreeding and outbreeding in humans. He concluded that some (but not all) crossbreeding of closely allied races could have relatively beneficial effects. The English and Scotch, for example, were the products of successive mixtures of closely related peoples and consequently are a highly variable population that produced genius (although also some "wretchedness"). The Irish, by contrast, were the product of close interbreeding of two "savage" tribes, and consequently the "Irish have hardly a single individual meriting a rank among the great names of history, or a contribution to literature, art, or science of first magnitude."[15]

The admixture of inferior stock (such as the Irish) and a superior stock (such as the English), according to East, led to a lowering of the superior one. East consequently had doubts about the wisdom of the United States' relatively open door policy for immigration from European countries. A melting pot, according to him, must "be sound at the beginning, for one does not improve the amalgam by putting in dross."[16] What about crosses among races he considered further apart? Here, East saw even greater potential problems, for such crosses could lead to a "breaking down of the inherent characteristics of each."[17] Although some positive recombinations might be possible, they seemed unlikely because human reproduction was not guided (as in agriculture). The fruit of generations of selection and adaptation could be squandered if unwise crossbreeding happened.[18]

East reflected the attitude of many geneticists in the early decades of the twentieth century who were influenced by the new Mendelian genetics, which was uncovering the complexity of inheritance. From experimental work, for

15. East, *Inbreeding and Outbreeding*, 257.

16. East, *Inbreeding and Outbreeding*, 264. This was of course the very opposite of the melting pot metaphor used by Israel Zangwill, the author who wrote the play, *The Melting Pot* (1909), that introduced the term into popular vocabulary.

17. East, *Inbreeding and Outbreeding*, 252.

18. This explanation allowed East, and others, to deal with Asians, who clearly had an impressive culture.

example, East knew that important traits could be the result of multiple genetic factors, and that these combinations of factors could be enhanced by
intensive breeding. More important for the issue of race mixing, East believed
that naturally occurring races contained complex sets of compatible physical
and mental traits. Human race mixing that involved individuals of "distant"
groups posed a danger because the positively linked traits that had been selected over long periods might break down in the offspring of a mixed cross,
with resulting loss of fitness.

The concerns voiced by Mjoen and East found expression in the most popular human genetics textbook of the 1930s, *Human Heredity* by Erwin Baur,
Eugen Fischer, and Fritz Lenz.[19] The authors reported that all human races
can interbreed, and that the resulting hybrids are fertile. Those races far apart
physically and mentally, however, do not mix well. Different races display different levels of achievement, they noted, and they repeated the suspicions that
mulattoes, and crosses between Lapps and Scandinavians, have constitutions
less able to withstand disease.[20]

On theoretical grounds, then, miscegenation did not appear to geneticists
to be a wise social policy. Although East claimed that the natural inferiority
of the Negro doomed him to extinction in the United States unless saved by
"amalgamation," he concluded that: "It seems an unnecessary accompaniment
to humane treatment, an illogical extension of altruism, however, to seek to
elevate the black race at the cost of lowering the white."[21]

The weight of scientific opinion was against race mixing, but this is not
to say that the argument was strong. What empirical evidence supported the
claims of the life-science community? The most extensive and widely quoted
study was that which had been undertaken in Jamaica by the geneticist and
eugenics advocate Charles Davenport.[22] In 1926 Davenport and Morris Steg-

19. Baur, Fischer, and Lenz, *Human Heredity* (New York: Macmillan, 1931). This is an English translation of the 1927 third edition of the original work (1921) published in German. The
authors involved in the translation added supplements and corrections. The work provided a
basis for later Nazi racial hygiene literature. See Robert Proctor, *Racial Hygiene Under the Nazis*
(Cambridge: Harvard University Press, 1988).

20. Baur, Fischer, and Lenz, *Human Heredity*, 175–84.

21. East, *Inbreeding and Outbreeding*, 254.

22. A modern biography has yet to be written of Davenport, although he is of obvious
interest and importance. Oscar Riddle's memoir is useful for an outline of Davenport's career
and bibliography: "Charles Davenport," *Biographical Memoirs, National Academy of Sciences of
the United States of America* 23 (1945): 75–110. Also, Mark Largent, *Breeding Contempt*, has a
good biographical discussion. All the major works listed earlier dealing with eugenics spend
considerable time on Davenport. In addition to Davenport's studies, a few others were routinely
cited on the biology of race mixing. The most common were by Eugen Fischer, who studied the

gerda, then a student from the University of Illinois's Department of Zoology, sailed to Jamaica on a research trip sponsored by the Carnegie Institution of Washington. Davenport chose the island because it contained an interesting racial composition: populations of what he considered pure white, pure Negro, and hybrids (mulatto). The trip had been made possible by a gift to the Carnegie Institution by an heir to a textile machinery fortune, the wealthy and eccentric eugenics enthusiast Wickliffe Preston Draper.[23]

The Jamaica study compiled hundreds of physical measurements and psychological test results on roughly 300 adults (a larger number of children were also observed but played a small role in the study). Although Davenport could point to some physical differences among his groups, he had little evidence of any biological dangers from race mixing. His main finding was that a few hybrids had long legs (Negro trait) and short arms (white trait). In the official Carnegie Institution report (1929), Davenport confessed, "We do not know whether the disharmony of long legs and short arms is a disadvantageous one for the individuals under consideration."[24] (In a popular report he indicated that those traits "would put them at a disadvantage in picking up things from the ground."[25])

The various psychological tests that Davenport conducted reinforced his concern with race mixing. He admitted that mulattoes display a wider range of variability than either pure line, but he claimed that they "comprise an ex-

---

Boer-Hottentot "hybrids" of South Africa, and Melville Herskovits, who studied the mulatto population in the United States. Fischer found no obvious biological problems, nor reduced fertility, in his group of subjects from mixed-race backgrounds. Fischer's work is problematic because he was not able to establish careful pedigrees for his subjects. See Eugen Fischer, *Die Rehobother Bastards und das Bastardierungsproblem beim Menschen* (Jena: Gustav Fischer, 1931). Herskovits studied "the American Negro" and claimed that the group was an amalgam, that is, most individuals who identified themselves as Negro were not "pure" but a mixture of Negro, white, and a small amount of Indian. Herskovits claimed that a blending of physical racial traits were seen in his data (rather than a Mendelian segregation of characteristics), but he did not explore the "superiority" or "inferiority" of these traits. See Melville Herskovits, *The American Negro* (New York: Alfred Knopf, 1928).

23. Draper, who died in 1972, supported a number of eugenics initiatives. He was the main benefactor of the Pioneer Fund (incorporated in 1937), whose first president was Harry Laughlin, longtime associate of Davenport. Draper later supported a number of segregationist causes. See William H. Tucker, *The Funding of Scientific Racism: Wickliffe Draper and the Pioneer Fund* (Urbana-Champaign: University of Illinois Press, 2002); Michael G. Kenny, "Toward a Racial Abyss: Eugenics, Wickliffe Draper, and the Origins of the Pioneer Fund," *Journal of the History of the Behavioral Sciences* 38 (2002): 259–83, and his, "A Question of Blood, Race, and Politics," *Journal of the History of Medicine* 61, no. 4 (2006): 456–91.

24. C. B. Davenport and Morris Steggerda, et al., *Race Crossing in Jamaica*, Carnegie Institution of Washington Publication No. 395(Washington, D.C.: Carnegie Institution, 1929), 471.

25. C. B. Davenport, "Race Crossing in Jamaica," *Scientific Monthly* 27, no. 3 (1928): 237.

TYPICAL MEN MEASURED AT MICO COLLEGE

FIG. 1—File No. 2615 Br.
FIG. 2—File No. 1520 Bl.
FIG. 3— File No. 2581 Bl.
FIG. 4—File No. 2712 Br.

Charles Davenport and His Jamaica Study

Charles Davenport supported the view that race mixing could be biologically harmful. Those who wished to argue against the advisability of interracial marriage often cited his writings. Davenport published his famous Jamaica study of race mixing in 1929. The pictures here illustrate some of the racial types he attempted to characterize.

■ From C. B. Davenport and Morris Steggerda, et al., *Race Crossing in Jamaica,* Carnegie Institution of Washington Publication No. 395 (Washington, D.C.: Carnegie Institution, 1929), 487.

ceptionally large number of persons who are poorer than the poorest of the Negroes or the poorest of the whites."[26] Although this is partly offset by hybrid populations that "contain persons better endowed in appreciation of music and in simple arithmetical or mental computations, as well as more resistant

26. Davenport, "Race Crossing in Jamaica," 238.

to certain groups of diseases, than a pure white population," Davenport sadly noted, "If only society had the force to eliminate the lower half of a hybrid population, then the remaining upper half of the hybrid population might be a clear advantage to the population as a whole, at least so far as physical and sensory accomplishments go."[27] Given the impracticality of such selection, he concluded that mixing would not benefit the population.

The vast bulk of Davenport's report focused on physical characteristics, and any but the most biased of readings would have to conclude that he found little evidence of any physical problems with race mixing. If Davenport's work represents the most extensive study, it is hardly surprising that some scientists concluded that insufficient evidence existed to warrant the conclusions drawn from theoretical considerations.[28] Yes, disharmonies might result; but did they? One of the most extensive reviews of the literature on race mixing, done shortly before Davenport published his study, stated the case well: "The problem is one of considerable difficulty and should inspire caution in the expression of opinion. Nevertheless, we constantly meet with verdicts delivered with much assurance and with a serene unconsciousness of the sources of error into which one may so easily fall. To separate the social from the biological effects of race mixture is a problem which few have attempted to solve, and which many, apparently, have not even considered."[29] Having an agnosticism regarding the biological ill effects of race mixing did not suggest to many life scientists that race mixing should be promoted or even tolerated. Most subscribed to the view of Samuel J. Holmes (the person who wrote the above quotation urging caution in our judgment) that race mixing constituted a "dangerous experiment."[30]

Charles Davenport believed the state had a right, and a responsibility, to regulate marriage in light of the knowledge gained from eugenics. From his Eugenics Record Office at Cold Spring Harbor, New York, he published a small booklet in 1913 that compiled the current state laws restricting marriage

27. Davenport, "Race Crossing in Jamaica," 238.

28. The few other studies done on race mixing, such as that of L. C. Dunn and A. M. Tozzer in Hawaii, or Harry Shapiro on Pitcairn Island, found no physical problems. Karl Pearson published a severe review of Davenport's Jamaica study, in which he argued that the only thing "that is apparent in the whole of this lengthy treatise is that the samples are too small and drawn from too heterogeneous a population to provide any trustworthy conclusions at all." Karl Pearson, "Race Crossing in Jamaica," *Nature* 126, no. 3177 (1930): 429.

29. Samuel J. Holmes, *A Bibliography of Eugenics* (University of California Publications in Zoology, vol. 25,1924), 465. This bibliography is a particularly valuable resource on the literature of the time.

30. Samuel J. Holmes, *Human Genetics and Its Social Import* (New York: McGraw Hill, 1936), 356.

and recommended future action.[31] Most of the existing laws had to do with limiting marriage because of either mental and physical conditions or consanguinity. But miscegenation occupied a portion of his book. He noted regretfully that the subject was one that "seems apt to arouse elemental passions" but stated that "no subject can be so threatening to the social order that it may not be fully discussed to the advantage of society."[32] Davenport went to pains to indicate that the opinions he expressed were not predicated on social bias, or an attempt to discriminate against any particular group. Rather, they were based on rational reflection and the best science of the day, with an eye toward the well-being of society.

After reviewing the various state laws, Davenport pointed out that in those states where the most race mixing had taken place, that is, the South, most state laws permitted a person with less than one-eighth "Negro blood" to marry a white. He then examined the problem from a scientific point of view, taking into account the positive traits (e.g., sense of humor, love of music, resistance to malaria) and negative ones (e.g., lack of appreciation of property distinctions, strong sex instinct without corresponding self-control) associated with the Negro race. Davenport couched his opposition to race mixing in terms of general eugenic principles. He noted, for example, that most racial characteristics are "separately inheritable" (i.e., they follow Mendel's law of independent assortment and are not linked), and that, therefore, in considering the results of crossing, we have to get past the issue of color: "Forget skin color and concentrate attention upon matters of real importance to organized society. Prevent those without sex-control or educability or resistance to serious disease from reproducing their kind."[33] In other words, Davenport opposed race mixing based on the same general principles that he opposed immigration of eastern Europeans: the genetic stock was inferior, and mixing it with the American race would threaten the future health of the country. The opposition was not, according to him, the result of prejudice, but of scientific deliberation.

The main eugenic issues facing the country differed in the North and South according to Davenport. The North had to contend with the rising tide of feeble minded and the threat of inferior immigrants. In the South, the influx of immigrants was less, but whites lived in close proximity to blacks and had a significant amount of race mixing (even if illegal). For Davenport, however,

31. Davenport, *State Laws Limiting Marriage Selection Examined in the Light of Eugenics,* Eugenics Record Office, Bulletin No. 9 (Cold Spring Harbor, N.Y.: Eugenics Record Office, 1913).
32. Davenport, *State Laws Limiting Marriage Selection,* 31.
33. Davenport, *State Laws Limiting Marriage Selection,* 36.

the problems of the North and South needed a common approach: "Both sections alike must not be content merely to bow their heads before the oncoming storm, but must take positive measures to increase the density of socially desirable traits in the next generation—by education, segregation and sterilization; and by keeping out immigrants who belong to defective strains."[34]

Davenport concluded that biological knowledge on miscegenation suggested the following recommendation: "No person having one-half part or more Negro blood shall be permitted to take a white person as spouse. Any person having less than half part, but not less than one-eighth part of Negro blood, shall not be given a license to marry a white person without a certificate from the State Eugenics Board."[35] So, in conformity with most of the southern laws, those with less than one-eighth Negro ancestry could marry a white. But, in those cases that fell between a person with one Negro parent and a person with one-eighth Negro ancestry, a State Eugenics Board would adjudicate, presumably on the basis of the balance of positive and negative traits exhibited by the person and his or her pedigree.

When Davenport published his pamphlet in 1913, twenty-nine of the forty-eight states of the Union had laws forbidding race mixing: nineteen forbade marriage between Negroes and whites; eight added "Mongolians" (Chinese and Japanese) to the ban; one specified Negroes as well as Croatan Indians; and one (Nevada) prohibited white persons from marrying persons of "Ethiopian, Malay [Filipino], Mongolian, or American Indian races."[36] Little changed until the period after World War II.[37]

One must, of course, place these "biological" opinions into their social context. Had biologists, like Davenport, been discussing the race mixing of Hawaiian birds, the question would not have been so highly charged. A series of issues fed into concern about race mixing, however. Although the supporters of eugenics could take some satisfaction from the immigration laws of the mid-twenties, which restricted the enormous influx of non-Anglo-Saxons,[38] the United States was experiencing the beginning of a dramatic internal mi-

34. Davenport, *State Laws Limiting Marriage Selection*, 36.
35. Davenport, *State Laws Limiting Marriage Selection*, 36.
36. Davenport, *State Laws Limiting Marriage Selection*, 30.
37. See Kennedy, *Interracial Intimacies,* for an interesting discussion of the changes in anti-miscegenation laws. Of considerable use also is Wallenstein, *Tell the Court I Love My Wife.*
38. The immigration act of 1924 restricted immigration from southern and eastern European nations, but not immigration within North America. Holmes, who lived in the West, was very concerned about the continuing influx of immigrants from Mexico. He was also concerned about the relative growth of the Negro population in the United States as a result of the restriction of European immigration. See his *The Negro's Struggle for Survival: A Study in Human Ecology* (Berkeley: University of California Press, 1937), 214–24.

TABLE III
LIMITS TO MARRIAGE BETWEEN RACES

	FORBIDDEN MARRIAGES.	STATUS OF MARRIAGE	MAXIMUM PENALTY.
ALABAMA	White person and Negro or *descendant* of a Negro to 3rd generation, inclusive, though one ancestor in each generation be white. Constitution forbids marriage of white person with Negro or descendant of Negro.		Imprisonment 2–7 yrs. for each party.
ARIZONA	Persons of Caucasian blood or their descendants with Negroes, Mongolians or their descendants.	Void	
ARKANSAS	Between a white and a Negro or mulatto.	Void	
CALIFORNIA	White person with Negro, mulatto, or Mongolian.	Void. No license to be issued.	
COLORADO	White person with Negro or mulatto, except in portion of state derived from Mexico.		Fine $500 or imprisonment, 2 yrs.
DELAWARE	White person with Negro or mulatto (as enrolled).	" Unlawful."	Fine $100 or imprisonment 30 days.
FLORIDA	White with a Negro (⅛ or more Negro blood). Constitution specifies persons of Negro descent to fourth generation, inclusive.	Null and void.	Imprisonment 10 yrs. or fine $1000.
GEORGIA	White persons with persons of African descent.	"Forever prohibited, null and void."	For officiating, fine, imprisonment 6 mo. and work in chain gang 12 mo.
IDAHO	White person with Negro or mulatto.	Illegal and void.	For solemnizing, fine $300 and imp. 3 mo.
INDIANA	White person with person having ⅛ or more Negro blood.	Void.	Imprisonment 10 yrs. and fine $100.
KENTUCKY	White person with Negro or mulatto.	Prohibited and void.	Fine $5000

State Laws on Race Mixing in the Early Twentieth Century

In a small book, Davenport listed all the state laws that restricted race mixing. Note in the first part, reproduced here, that different states had different laws, and that the definition of *race* also varied from place to place.

■ State laws restricting interracial marriage. Charles Davenport, *State Laws Limiting Marriage Selection Examined in the Light of Eugenics,* Eugenics Record Office, Bulletin No. 9 (Cold Spring Harbor, N.Y.: Eugenics Record Office, 1913), 28.

gration: southern blacks moving to the North.[39] During the second half of the twentieth century, race no longer remained primarily a southern issue but arose as a challenge in most urban centers of the country. Southern opposition to race mixing rested on the legacy of slavery, concern with the property rights of whites, and political desires to maintain white dominance. Other issues were added to these in the North. Since race mixing was outside the law, the fruit of such unions (mulattoes) were associated with children born out of wedlock, or the children of mistresses and prostitutes.[40] Intelligence tests administered to military men during the First World War contributed to the discussion of race mixing. Negroes' scores were on average below those of whites, although a significant percentage of northern blacks scored higher than southern whites. The tests, although of wide interest, could be interpreted in different ways. It wasn't clear if economic and social factors contributed to the scores, nor was it clear what sort of intelligence was being tested. Superficially, however, the scores were interpreted to support the deeply held view that Negroes were intellectually inferior to whites (even the Irish).[41]

39. For a discussion of the black migration, see Carole Marks, *Farewell—We're Good and Gone: The Great Black Migration* (Bloomington: Indiana University Press, 1989); Nicholas Lemann, *The Promised Land: The Great Black Migration and How It Changed America* (New York: Alfred Knopf, 1991); and Florette Henri, *Black Migration: Movement North, 1900–1920* (Garden City, N.Y.: Doubleday, 1976).

40. Holmes, *The Negro's Struggle for Survival*, 174–75.

41. For background to the intelligence test issue, see Cravens, *The Triumph of Evolution*.

3 Challenges to Opinions on Race Mixing

During the 1930s, a major shift in attitudes toward race mixing occurred within the scientific community, and by the 1940s, a significant reversal of opinion on the allegedly negative biological consequences of miscegenation took place. Various influences contributed to this shift.[1] A major factor emerged as part of the strong reaction in the United States to the emerging National Socialist Racial Hygiene program in Germany. Although German geneticists and physicians maintained close contacts with American scientists, the extreme Aryan supremacy literature and the Nuremberg Laws in Germany alarmed many Americans. It was one thing to enact immigration policies that sought to maintain an Anglo-Saxon majority in the United States, but the German laws aimed at Jews and other "inferior white" races passed the limits of what was acceptable in a democratic society. To be sure, some of the American and British reaction against German racial hygiene continued to contain implicit racist notions, but the net result strengthened the move by geneticists and anthropologists in questioning the abuse of racial categories.[2]

The Boas School

The rise of the Boas School of cultural anthropology was another important element in the altering opinion. Franz Boas, at Columbia University in New York, and his students, stressed the importance of the environment and of the cultural context in which people lived rather than biological determinants. They also challenged the assumption that "racial characteristics" relied primarily on hereditary factors and contested the view that race mixing would produce inferior individuals.[3] The early development of cultural anthropology

1. See references in chapter 2, footnote 3.

2. For an interesting, and disturbing, discussion of the continuing implicit racism in much of the humanitarian and liberal literature on genetics, see Mikuláš Teich, "The Unmastered Past of Human Genetics," in Mikuláš Teich and Roy Porter, *Fin de Siècle and Its Legacy* (Cambridge, Cambridge University Press, 1990), 296–324.

3. Boas's students were energetic in their rejection of scientific racism and wrote extensively against the assumptions of black racial inferiority and against the alleged biological problems of

took place contemporaneously with, and in conscious opposition to, much of the eugenics movement and the scientific racism associated with it. Boas and his "school" played an important part in the move away from the scientific support of anti-miscegenation during this period, but we should not forget that Boas himself had been fighting the battle for a long time.

Boas had experienced anti-Semitism in Germany before moving to the United States in 1887, and his life in America was characterized by a firm commitment to social liberalism and the fight against racial prejudice. His early ethnographic research on Eskimos, furthermore, instilled in him a deep respect for "primitive people." Boas combated racism for decades, and he had an enormous impact on American life. Much of his effort was aimed at erasing the discrimination directed against the wave of immigrants entering the country from southern and eastern Europe. Part of his argument consisted of exposing the dubious nature of racial purity. He noted that European peoples had been mixing for centuries, and that any reference to pure "stock" was imaginary. In *The Mind of Primitive Man* (1911), he wrote, "The history of Europe proves that there has been no racial purity anywhere for exceedingly long periods, neither has the continued intermixture of European types shown any degrading effect upon any of the European nationalities."[4] Boas believed that environmental influences altered physical characteristics that distinguished different European "races," and that European populations, rather than conforming to a generalized standard, displayed enormous variation. He also stressed that little evidence existed to support the claims of those critical of the "mongolization" of America, or of those who, like himself, welcomed it. He designed research to demonstrate the environmental influence on the offspring of immigrants, and he encouraged his students to study the alleged effects of race mixing.

Although much of Boas's focus centered on European immigrants, his work had implications for more distant racial groups, and he extended his discussion to cover the mixing of Negroes with Euro-Americans. In his American Association for the Advancement of Science Presidential Address of 1931, he strongly opposed the views of those like Davenport who argued that race mixing should be discouraged. After repeating that European groups had intermingled for years and that all populations contain extensive variation, he specifically addressed the issue of the mixing of distant races, that is, African and

---

race mixing. Most well known are Alfred Kroeber, Melville Herskovits, Ruth Benedict, Margaret Mead, Ashley Montagu.

4. Franz Boas, *The Mind of Primitive Man* (New York: Macmillan Company, 1911), 260.

European. Little direct evidence for any position existed, and he was forced to admit that it was not "easy to give absolutely conclusive evidence in regard to this question."[5] However, he pointed out that in the few cases that had been studied, "there does not seem to be any reason to assume unfavorable results, either in the first or in later generations of offspring."[6] And, flying in the face of much public opinion, he stated, "The biological observations on our North American mulattoes do not convince us that there is any deleterious effect of race mixture so far as it is evident in anatomical form and function."[7] He went on to discuss the influence of the environment on individuals and the influence of culture on behavior, and then summarized his position forcefully:

> I believe the present state of our knowledge justifies us in saying, that while individuals differ, biological differences between races are small. There is no reason to believe that one race is by nature so much more intelligent, endowed with great will power, or emotionally more stable than another, that the difference would materially influence its culture. Nor is there any good reason to believe that the differences between races are so great, that the descendants of mixed marriages would be inferior to their parents. Biologically there is neither good reason to object to fairly close inbreeding in healthy groups, nor to intermingling of the principal ones.[8]

Boas did not ignore the importance of social forces. He recognized that social stratification into racially based social groups would lead to discrimination and resistance to mixing. He saw it as a major problem in American society, one not likely to be solved in the near term. But he located the problem in the social sphere, not in the biological.

Columbia University was the institutional base for Boas, and his students there contributed substantially to furthering his goals. Melville Herskovits spent years studying the hereditary results of race mixing. He was among the first to argue that the "American Negro" was largely a product of mixture. In doing so, he was contesting the results of the 1920 census that reported only 15.9 percent of Negroes in the United States were of mixed blood.[9] Unlike

5. Franz Boas, "Race and Progress," *Science*, 74, no. 1905 (1931): 3.

6. Boas, "Race and Progress," 3. Also see chapter 2, footnote 22.

7. Boas, "Race and Progress," 3.

8. Boas, "Race and Progress," 6.

9. Melville Herskovits, *The American Negro*, 5. On Herskovits, see George Eaton Simpson, *Melville J. Herskovits* (New York: Columbia University Press, 1973). Herskovits went on to become one of the leading figures in what is now called African Studies. He maintained an interest in race and worked to show the sophisticated nature of African history and culture, in contrast to the popular view that Africa was characterized by a lack of cultural achievement.

World War II and Scientific Racism

Ashley Montagu and other American scientists were deeply concerned about the extreme nature of the literature on race in Nazi Germany and the implementation of laws based on it. Until the late 1920s, much of the German eugenics literature had not differed substantially from that of the literature on eugenics in the United States. Increasingly, however, the German eugenics movement merged with the more extreme views of Nordic supremacy, a movement that considered Nordic peoples a "master race," superior to other European groups and even more superior to non-European races. The view that the peoples of northern Europe were superior to those of other parts of the globe was not new or confined to Germany. The position was widely held by American eugenicists and informed their support for immigration restrictions that were enacted by Congress in the 1920s. What made the Nordic supremacy objectionable was the political direction it took under the Nazi government.

In 1935, the Nazi government decreed that all Jews (even someone half Jewish or a quarter Jewish) were no longer citizens of the country and no longer could vote. Jews were dismissed from all government positions. The same year, marriage between Jews and non-Jewish Germans was prohibited, as was extramarital sexual relations between Jews and Germans. Jews were no longer allowed to employ female citizens under the age of 45 as domestic help. Punishment for violating the marital laws or engaging in prohibited sexual relations was hard labor or prison. The following year, Jews were excluded from all professional positions, including medical, educational, and administrative jobs in industry. The propaganda slide above, titled "Race Defilement," depicts "mixed" couples (Aryan and Jewish). The caption reads "Women and Girls, the Jews are your ruin!"

The crudeness of the German laws made many Americans reflect on the anti-miscegenation laws in the United States and on the racism that infused much of the American eugenics literature.

■ Nazi slide entitled "Race Defilement" from a lecture by the leader of the S.S., the chief of the Race and Settlement Main Office, 1936. Courtesy of the United States Holocaust Memorial Museum.

Davenport's Jamaica study, Herskovits measured thousands of subjects, collected data at distant locations, and acquired genealogical information as well. He came to a number of unexpected conclusions. The most interesting were that only a little over 20 percent of American Negroes were unmixed, and that the American Negro seemed to exhibit less variability than the "unmixed" parent populations.[10] This latter point was especially surprising. Not only did his sample show a reduced comparative variability, but the variability within Negro families was also less. Herskovits believed that a new physical type was coming into existence in the American Negro, most likely as a result of decrease in the amount of race mixing between whites and Negroes. Since geneticists and breeders had considered decreased variability as an indicator of purity, Herskovits's study, which focused upon a clearly mixed group, presented an anomaly. Herskovits suggested that the conventional notion of the major races as pure had to be rethought; that we needed to look at the actual history of groups to determine how mixed or homogeneous they were and not assume that all groups were largely homogeneous.[11] More radical, Herskovits wrote that his study showed how "little we are able to define a word that has played such an important role in our political and social life, while it further illustrates how much we take for granted in the field of the genetic analysis of human populations."[12] This ignorance about the biological meaning of race became one of the principal points of issue for the students of Boas and was used to undercut many of the myths surrounding the discussion of race and, by extension, race mixing.

The Boas School campaigned vigorously against the scientific racism that had characterized much of anthropology and genetics in the twenties and early thirties. They argued for a more relativistic approach to race that stressed environmental as well as hereditary factors. Among the most prominent of this group were Margaret Mead, Ruth Benedict, Ashley Montagu, and Otto Klineberg, who joined Herskovits to encourage a different perspective on race. Along with Herskovits, Klineberg and Montagu gave considerable attention to the issue of race mixing.[13]

10. Herskovits, *American Negro*, 5. See also, Melville Herskovits, *The Anthropometry of the American Negro* (New York, Columbia University Press, 1930).

11. Herskovits, *American Negro*, 68–72.

12. Herskovits, *American Negro*, 82.

13. Earnest Hooton, the leading physical anthropologist in the United States, joined the cultural anthropologists in distancing himself from Nazi racial arguments. In 1935 he summarized the studies that had been done on miscegenation and stated that they demonstrated no ill biological effects resulting from such crosses. See Hooton, "Homo Sapiens—Whence and Whither," *Science* 82, no. 2115 (1935): 19–31.

In his 1935 popular book, *Race Differences,* Otto Klineberg directly addressed the scientific literature that warned of the dangers of race mixing. Drawing on numerous critiques of anti–race-mixing literature, such as those of William Castle, he reviewed the writings of Davenport and Mjoen and pointed out their weaknesses. To this, he added a discussion of mental traits of "mixed bloods." Klineberg had been interested in the alleged mental inferiority of Negroes that grew out of the First World War studies of recruits. In 1944, he edited and contributed to a volume, *Characteristics of the American Negro,* that focused extensively on the issue of Negro intellectual ability. Making use of the work of several researchers, such as Thomas Garth, Klineberg concluded that the very notion of ranking races on their mental or psychological characteristics lacked any scientific evidence, and that competent scholars in the area generally agreed that it was not a valid endeavor.[14]

What implications did these studies hold for ideas on race mixing? Klineberg clearly and unambiguously argued that there is no reason to treat individuals

> differently because they differ in their physical type . . . There is no reason to pass laws against miscegenation. The human race is one, biologically speaking there are no sub-varieties whose genes are mutually incompatible, or whose crossing will necessarily lead to degeneration. Race mixture is not in itself harmful . . . If two individuals of different stock wish to marry, any objection by the state is an unwarranted interference in a matter which concerns them alone, and which in any case has not been shown to have any harmful consequences. Laws directed against mixture in order to maintain race purity have no meaning, since every large population in the world already contains within it a varied assortment of physical types.[15]

Ashley Montagu

Although the Boas School as a whole made efforts to be vocal and publicly active on issues of race, perhaps the most strident adversary of scientific racism and its opposition to race mixing was Ashley Montagu. Born in London, Montagu had studied anthropology and psychology in England before moving to the United States in 1931, where he directed the Division of Child Growth and Development at New York University for seven years.[16] While in New

14. Otto Klineberg, ed., *Characteristics of the American Negro* (New York: Harper and Brothers, 1944), 95.

15. Otto Klineberg, *Race Differences* (New York: Harper and Brothers, 1935), 345–47.

16. Montagu had an atypical career. After NYU he obtained a position in the Department of Anatomy at the Hahnemann Medical School in Philadelphia. Later, he was chair of Anthro-

York, he enrolled in a Ph.D. program at Columbia and received his doctor-
ate in anthropology in 1937. Like Margaret Mead, another of Boas's students,
he devoted much of his career to popularizing anthropology and relating its
research to contemporary issues.

Shortly after completing his Ph.D., Montagu began what turned out to be
an extended critique of the use of the term *race*. He started with a set of papers,
and later revised them into a short book, which continued to be revised and
expanded into a quite large and famous volume. From the beginning, he used
two arguments that became his standard two-punch assault. The first was a
historical sketch of how the concept of race had changed since its first incep-
tion in the eighteenth century. His 1941 article, "The Concept of Race in the
Human Species in Light of Genetics,"[17] attributed the first scientific use of the
term *race* to the early writings of the famous Enlightenment naturalist Georges
Louis Leclerc, comte de Buffon. Montagu noted that Buffon regarded his six
races as a "convenience"; that is, Buffon could have divided them differently.[18]
Montagu also discussed the work of Johann Friedrich Blumenbach whose *De
Generis Humani Varietate Nativa* was widely quoted on racial categories, and
who like Buffon had stated that his classification was to be seen as a convenience and that it did not undercut the basic unity of the human species.

So, the early history of the term *race* revealed its somewhat arbitrary nature
and showed that it had not originally been proposed as a fixed entity. Montagu's second point was that later anthropologists had accepted an altered sense
of race, one that had its roots in Aristotelian philosophy, in the theological
doctrine of special creation, and in the early nineteenth-century classification
system of Cuvier. He meant by this that nineteenth-century anthropology
used outmoded concepts like the Aristotelian view that species possessed a set
of essential characteristics, and the theological belief that considered species
individual acts of Creation, and combined these ideas with the static anatomy
of Cuvier, which argued that species were unchangeable. The resulting concept

---

pology at Rutgers, from which he resigned and spent most of the rest of his life in Princeton,
New Jersey.

17. Ashley Montagu, "The Concept of Race in the Human Species in Light of Genetics,"
Journal of Heredity 32, no. 8 (1941): 243–47. Also see his article "The Genetical Theory of Race,
and Anthropological Method," *American Anthropologist* 44 n.s. (1942): 369–75. This latter paper
was read at the annual meeting of the American Association of Physical Anthropologists in
1940.

18. Buffon's legacy was more ambiguous than Montagu's description implies. For an excellent
discussion, see the classic article by Philip R. Sloan, "The Idea of Racial Degeneracy in Buffon's
Histoire Naturelle," *Studies in Eighteenth-Century Culture* 3 (1973): 293–321.

of race assumed that races were fixed entities that could be rigorously defined by their anatomical features. Even the Darwinian revolution, according to Montagu, had not substantially altered this view, for naturalists assumed that species (and by extension races) retained their physical characteristics *unless* natural selection operated to alter them. Anthropologists, furthermore, considered human evolution recent and the product of a relatively short time span, so that racial change was not relevant for comparing contemporary races.

Montagu argued that anthropologists had taken an eighteenth-century concept, one formulated for convenience, and had mistakenly assumed it had a fixed and definable reality. He attacked the common practice in physical anthropology of tabulating physical measurements to construct a list of individual traits that defined specific races. Satirically, he called this the "omelette conception of race": "The process of averaging the characteristics of a given group, knocking the individuals together, giving them a good stirring, and then serving the resulting omelette as a 'race' is essentially the anthropological process of race-making. It may be good cooking, but it is not science, since it serves to confuse rather than to clarify . . . The omelette called 'race' has no existence outside the statistical frying-pan in which it has been reduced by the heat of the anthropological imagination."[19]

What did he mean? A long tradition in physical anthropology had attempted to define the different human races by averaging measurements from individuals. This process grew out of the methods used in natural history to define the physical characteristics of different species. Naturalists collected series of specimens and measured, recorded, and averaged their characteristics to give a "typical" set of traits. In those species that were known to have geographical varieties, naturalists employed a similar method: specimens from known locales were collected and examined to yield a set of traits that differed in some specified ways from the larger, more general population. An entire body of literature in natural history discussed the details of such classification, and anthropologists believed that they were on solid ground applying it to humans. But there was a problem, one that Montagu focused on and used to undermine the approach to race that anthropologists had accepted.

Research done in population genetics was transforming how scientists in the 1940s understood evolution. Montagu knew, and well understood, how scientists like Theodosius Dobzhansky were combining Mendelian genetics with field studies to "capture" evolution in action. Genetics (by which he

19. Montagu, "The Concept of Race," 245.

generally meant the population genetics of Dobzhansky and others) put the practice of classification, and especially classification of races, in a new light. As early as April 1940, Montagu gave a paper at the annual meeting of the American Association of Physical Anthropologists, where he laid out how the new genetics altered the meaning of race as used by anthropologists, that is, the attempt to define a set of metrics that characterized the different races. *Race* for Montagu had a new biological meaning. The human species consisted of a large set of populations, the descendants of some original ancestral population. In time, the ancestral population had dispersed and had become geographically separated into multiple populations, which were subject to random variations and had come to vary in their gene frequencies. Moreover, the individuals in these populations were also subject to gene mutations, at different rates and in different characteristics. What Montagu initially stressed was the dynamic aspect of race: populations are in constant flux, and what we call race is really just a population with a particular set of gene frequencies. These frequencies are shifting, and they are operated on by secondary forces: migration, social and sexual selection, endogamy, exogamy, and so forth.[20] So, there are two separate sets of forces molding human populations: variability in local groups and the social forces of mixing. The then current definitions of race were woefully inadequate since they focused on "arbitrary and superficial selection of external characters."[21] Montagu concluded with a proposed new definition and a new terminology. Since *race* appeared to him a somewhat incoherent concept, he chose to use the term *ethnic group*, a concept earlier used by Julian Huxley and A. C. Haddon in their widely read book *We Europeans: A Survey of "Racial" Problems* (1936).

Huxley, the prime author, was arguing against the racial theories that were emerging from Hitler's Germany, which stressed the racial purity of northern Europeans (Aryans) and the inferiority of other European groups. In contrast, Huxley emphasized the inherently mixed nature of all European races. The biological concept of race, according to Huxley, could not be applied to groups such as the Germans, the French, and so forth. He suggested *ethnic group* for these mixed populations. Huxley did not confine his treatment to just Europeans. All the groups recognized by anthropologists are mixtures of peoples, so no pure races existed anywhere at the present.[22] Consequently, the social

---

20. Montagu, "The Genetical Theory," 374.

21. Montagu, "The Genetical Theory," 374.

22. Huxley thought that some "primary" races might have hypothetically existed in the distant past, but we have no direct knowledge of them.

policy that forbids race mixing "turns out not to be primarily a matter of 'race' at all, but a matter of nationality, class or social status."[23] Huxley concluded that "Racialism is a myth, and a dangerous myth at that. It is a cloak for selfish economic aims which in their uncloaked nakedness would look ugly enough. And it is not scientifically grounded."[24]

Montagu picked up Huxley's skepticism about the literature on race and his suggestion that ethnic group replace the former concept, and most important, he stressed the fundamental genetic unity of humans. Ethnic groups are merely populations of *Homo sapiens* that have different relative frequencies of some genes, and these differences are maintained (or not maintained) by physical and social barriers. If Negroes and whites were "freely permitted to marry, the physical differences between Negroes and whites would eventually be completely eliminated through the more or less equal scattering of their genes though the population."[25]

In the first edition (1942) of his best known book, *Man's Most Dangerous Myth: The Fallacy of Race*, Montagu brought together (with some rewriting and revision) a number of articles he had published on race. He extended Huxley's warning that racialism was a dangerous myth to the broader claim that the concept of race itself was a myth. Montagu's book went through six editions and reflected an ever more strident stand on the issue. The first edition admits that there is a biological meaning of race: "Mankind is comprised of many groups which are often physically sufficiently distinguishable from one another to justify their being classified as separate races."[26] These are very large groups—Mongolian, Caucasian, Negro, Australo-Melanesian, and Polynesian. Races, he reminds us, are temporary mixtures that are in flux and, as experience shows, among which there is complete interfertility. Montagu argued that the genetic differences among these groups are slight; that is, they differ in the distribution of a small number of genes. More important, he went on to state that the many complex groups to which we normally apply

23. Julian Huxley and A. C. Haddon, *We Europeans: A Survey of "Racial" Problems* (New York, Harper and Brothers, 1936), 232.

24. Huxley and Haddon, *We Europeans*, 232. A perhaps not surprising, but ironic, result of the reaction against Nazi racial science, which stressed the superiority of Nordic types, was the narrowing of the concept of race from a concept that applied to many groups we would today consider nationalities (e.g., French, Irish) to one that focused primarily on color. See Matthew Frye Jacobson, *Whiteness of a Different Color: European Immigrants and the Alchemy of Race* (Cambridge: Harvard University Press, 1998).

25. Montagu, "The Genetical Theory," 375.

26. M. F. Ashley Montagu, *Man's Most Dangerous Myth: The Fallacy of Race* (New York: Columbia University Press, 1942), 4.

the term *race* are the products of cultural as well as physical difference, and that the mixture among these groups is considerable. In this sense, *race* has lost its anthropological and biological significance. Race mixing was a major topic in Montagu's book, and in a long chapter, he explored many dimensions of the issue. He made it clear that objections to race mixing are, basically, reflections of social factors, not biological ones, and that, "There can be little doubt that those who deliver themselves of unfavorable judgments concerning race-crossing are merely expressing their prejudices."[27] Taking the offensive, he moved the discussion to the general issue of hybridization. Montagu briefly reviewed the importance of crossbreeding and outbreeding in agriculture and then pointed out the irony of attempting to argue against race mixing by reference to hybridization research:

> It is, indeed, a sad commentary upon the present condition of western man that when it is a matter of supporting his prejudices he will so distort the facts concerning hybridization as to cause laws to be instituted making it an offense against the state. But when it comes to making a financial profit out of the scientifically established facts, he will employ geneticists to discover the best means of producing hybrid vigor in order to increase the yield of some commercially exploitable plant or animal product. But should such a geneticist translate his scientific knowledge to the increase of his own happiness and the well-being of his future offspring, by marrying a woman of another color or ethnic group, the probability is that he will be promptly discharged by his employer.[28]

Montagu most likely had in mind the racial hygiene laws in Germany when he wrote this in 1942, but he certainly also intended the critique to apply to U.S. anti-miscegenation laws and earlier geneticists like East, who stressed the value of outbreeding in hogs and corn but cautioned against it in humans.

In his book, Montagu provided a review of the existing studies that focused on race mixing and of their interpretations. One of the points he attempted to establish using this literature is how much the social place of crossbreeds affects their well-being and achievement. The issue is not heredity, but culture. In those places where the offspring of a mixed-race couple are not discriminated against, they do as well as nonmixed offspring. His review, in addition, noted that physically there is even the suggestion of hybrid vigor in some cases. Montagu ridiculed Davenport and Steggerda's figures, showing an alleged disharmony in leg and arm length in Jamaican "Browns," and stated unequivo-

27. Montagu, *Man's Most Dangerous Myth*, 99.
28. Montagu, *Man's Most Dangerous Myth*, 102–103.

cally, "The whole notion of disharmony as a result of ethnic crossing is a pure myth . . . the differences between human groups are not extreme enough to be capable of producing any disharmonies whatever."[29]

Montagu concluded the first edition of *Man's Most Dangerous Myth* with an appendix, "State Legislation Against Mixed Marriages in the United States." The data is from Chester Vernier's *American Family Laws*,[30] and Montagu quoted Vernier's remarks on the data: "Such legislation is not based primarily upon physiological, psychological, or other scientific bases, but is for the most part the product of local prejudice and of local effort to protect the social and economic standards of the white race."[31] Montagu, a man always ahead of his time, also noted that the anti-miscegenation laws contravened the United States Constitution, but that the Supreme Court had never handed down a decision that related to them.

The UNESCO Statement

Subsequent editions of *Man's Most Dangerous Myth* expanded the basic arguments about race and race mixing and grew from the first edition's 214 pages to 699 in the 1997 sixth, and final, edition.[32] The book had a large impact and is still in print. Montagu also had an important role in crafting the UNESCO statement on race. In response to a resolution passed by the fourth general conference of UNESCO as part of its campaign against racism, the director-general charged the Social Sciences Department to gather scientific information on the problems of race, disseminate it, and prepare an educational campaign.[33] The head of the department, the Brazilian anthropologist Arthur Ramos, called together a set of social scientists to draft a statement on race. Although Ramos soon died, the project went forward, and a committee began meeting at the end of 1949 in Paris. Montagu was *rapporteur* (reporter) of the committee. After a few days, they completed a draft and circulated it among a group of social and life scientists, which included Julian Huxley (at the time UNESCO's director-general), Theodosius Dobzhansky, Otto Klineberg, and

29. Montagu, *Man's Most Dangerous Myth*, 119.

30. Davenport published an earlier tabulation of state laws; see Charles Davenport, *State Laws Limiting Marriage Selection*.

31. Montagu, *Man's Most Dangerous Myth*, 188.

32. An interesting research project would be to compare the various editions and show how they map onto the social issues of the day. The first edition has an appendix of miscegenation laws that does not appear later since the issue was moot given *Loving v. Virginia* in 1967.

33. *Records of the General Conference of the United Nations Educational, Scientific and Cultural Organization, Fourth Session* (Paris: UNESCO, 1949), 22.

TEXT OF THE STATEMENT
ISSUED 18 JULY 1950

1. Scientists have reached general agreements in recognizing that mankind is one : that all men belong to the same species, *Homo sapiens*. It is further generally agreed among scientists that all men are probably derived from the same common stock; and that such differences as exist between different groups of mankind are due to the operation of evolutionary factors of differentiation such as isolation, the drift and random fixation of the material particles which control heredity (the genes), changes in the structure of these particles, hybridization, and natural selection. In these ways groups have arisen of varying stability and degree of differentiation which have been classified in different ways for different purposes.

13. With respect to race-mixture, the evidence points unequivocally to the fact that this has been going on from the earliest times. Indeed, one of the chief processes of race-formation and race-extinction or absorption is by means of hybridization between races or ethnic groups. Furthermore, no convincing evidence has been adduced that race-mixture of itself produces biologically bad effects. Statements that human hybrids frequently show undesirable traits, both physically and mentally, physical disharmonies and mental degeneracies, are not supported by the facts. There is, therefore, no *biological* justification for prohibiting intermarriage between persons of different ethnic groups.

The UNESCO Statement on Race

Ashley Montagu, L. C. Dunn, and Theodosius Dobzhansky were outspoken in their criticism of scientific racism. After the war, they were involved in drafting a statement on race for the United Nations Educational, Scientific, and Cultural Organization (UNESCO). The first statement (of what became a number of statements), published in 1950, contained this comment on race mixing.

■ Portion of the UNESCO statement that refers to race mixing. From *The Race Question*, UNESCO Publication 791 (Paris: UNESCO, 1950).

Gunnar Myrdal. A revised draft was sent for comments to some of the same individuals plus additional experts, such as Edwin Conklin, L. C. Dunn, Curt Stern, and H. J. Muller, and a final draft was produced by Ashley Montagu.

The UNESCO statement defines a human race as a population that may vary from other populations (due to various isolating factors in the past) and clearly distinguishes it from the everyday use of the term *race,* which refers commonly to groups based on national, religious, linguistic, or cultural factors. The statement also asserts that races (in either sense) have been mixing from earliest times, and that "no convincing evidence has been adduced that race mixture of itself produces biologically bad effects. Claims that human hybrids frequently show undesirable traits, both physically and mentally, physical disharmonies and mental degeneracies, are not supported by the facts. There is, therefore, no biological justification for prohibiting intermarriage between persons of different ethnic groups."[34]

The UNESCO statement on race immediately drew criticism, in part because those who drafted it came primarily from the social sciences. In response, a new committee with the geneticist L.C. Dunn as *rapporteur* prepared a second statement on race.[35] The new committee, consisting of physical anthropologists and geneticists, repeated, with more authority, the sentiment that "There is no evidence that race mixture produces disadvantageous results from a biological point of view." Attitudes towards race mixing, the statement went on to say, "whether for good or ill, can generally be traced to social factors."[36]

Although Montagu's campaign to replace the term *race* with *ethnic group* continued to be controversial within the scientific community, his writings

34. Ashley Montagu, *Statement on Race,* 3rd ed. (Oxford: Oxford University Press, 1972), 10.

35. For example, an editorial remark in the *American Journal of Physical Anthropology* in 1951 (6, no. 1, p. 3) stated that professional anthropologists were astonished to be "by-passed" in the selection of the UNESCO panel. Some of the objections were incorporated into *The Race Concept: Results of an Inquiry* (Paris: UNESCO, 1951). On Dunn, see Melinda Gormley, "Geneticist L. C. Dunn: Politics, Activism, and Community" (Ph.D. diss., Oregon State University, 2007).

36. Montagu, *Statement on Race,* 146. Later UNESCO statements (1964, 1967) make the same points. The first statement had been criticized not only because it was written primarily by social scientists but also because of the strongly stated universal brotherhood theme that infused it (largely Montagu's influence). See Elazar Barkan, "The Politics of the Science of Race: Ashley Montagu and UNESCO's Anti-racist Declarations," in Larry T. Reynolds and Leonard Liebermann, eds., *Race and Other Misadventures* (Dix Hills, N.Y.: General Hall, 1996), 96–105, and Michelle Brattain, "Race, Racism, and Antiracism: UNESCO and the Politics of Presenting Science to the Postwar Public," *American Historical Review* 112, no. 5 (2007): 1386–1413. Brattain notes that opposition also came from some geneticists who disagreed with Montagu's desire to eliminate the biological meaning of race, or who continued to hold different views on race. She demonstrates that the UNESCO project did not succeed in altering public opinion to the extent envisioned by Montagu and others. However, the document continued to be used to combat anti–race-mixing opinions. Provine, "Geneticists and the Biology of Race Crossing," also discusses some of the early opposition by geneticists to the UNESCO statement. For the genetics that informed the statement, see Paul Farber, "Changes in Scientific Opinion on Race-mixing: The Impact of the Modern Synthesis," in Paul Farber and Hamilton Cravens, eds., *Race and Science: Scientific Challenges to Racism in Modern America* (Corvallis: Oregon State University Press, 2009), 130–51.

and his influence on the formulation of the UNESCO statement on race provided key tools for undercutting the scientific arguments against race mixing.
Montagu's arguments rested on research that was being done in genetics, and
he made full use of the writings of population geneticists such as Theodosius
Dobzhansky. Of critical importance in the attempts to shift attitudes on race
mixing, therefore, were the changes that took place in the life sciences, particularly in the new genetics involved in transforming the modern theory of evolution. What exactly was this new science and how did it relate to the issue?

4 The Modern Synthesis

Boas and his students led the struggle against scientific racism and anti-miscegenation opinion. Less well known are the contributions of life scientists to the new understanding of race and race mixing. Many of these scientists strongly expressed their opposition to scientific racism, but the role of their science has not been adequately appreciated in histories of the opinion shift about race and race mixing.[1] In part, this is because in the thirties and forties, no new extensive *empirical* studies contradicted the earlier view that race mixing represented a "dangerous experiment." To overlook the contribution of the life sciences, however, is to miss an important component of the story. Starting in the late thirties, a set of *theoretical* works on the theory of evolution transformed biology and with that the concept of race. The *modern synthesis*, as it has come to be called, reestablished a Darwinian perspective on evolution and constructed a solid foundation in population genetics on which to build. By the fifties, the modern synthesis had swept the field, and it has continued to dominate the life sciences. The success of the modern synthesis has had broad social implications, for since its earliest days, many of its architects and supporters were public intellectuals who made a point of arguing for what they took to be its social significance. Even before the synthesis attained hegemony in the life sciences, it exerted significant social influence: the foundation of Huxley's and Montagu's arguments was based on the new genetics and the evolutionary perspective of the modern synthesis. By the 1960s, the new evolutionary theory informed how the educated public understood race and, more important for our story, race mixing.

1. Generally, historical studies claim that this change in ideas resulted from humanitarian or ideological causes, not from scientific considerations. See, for example, Provine, "Geneticists and the Biology of Race Crossing." Barkan, *The Retreat of Scientific Racism*, and Joseph Graves, *The Emperor's New Clothes: Biological Theories of Race at the Millennium* (New Brunswick, N.J.: Rutgers University Press, 2001).

Dobzhansky and the Concept of Race

Race was a central concept used in constructing the modern synthesis, and consequently the theory held important implications for evaluating the biological aspect of race mixing. We can see this clearly in the work of Theodosius Dobzhansky, who in 1937 published the first monograph setting out the newly emerging theory of evolution, *Genetics and the Origin of Species*.[2] Dobzhansky began his career working on ladybird beetles and developed an interest in the variation within populations of the same species. His scientific goals, even at this early stage, transcended the desire to sort out problems in insect classifications. Instead, he sought to understand the process of evolution and its implications for humans. After coming from the Soviet Union to the United States in 1927 to work in Thomas Hunt Morgan's lab, Dobzhansky focused many of his studies on the genetic differences in populations of *Drosophila pseudoobscura*, a fruit fly inhabiting the southern and western parts of the United States and extending down into Mexico.[3] Dobzhansky found this species attractive because of the work of Donald Lancefield (a former student of Morgan's), who in 1929 described two "races" of the species. They were of interest because although the races appeared anatomically the same, when individuals from the different races were crossed, they produced sterile males (but fertile females). Since the inability to interbreed successfully is often a sign that the individuals come from different species, this infertility of the male offspring was especially intriguing. Luckily, Dobzhansky had a student from the Seattle area, where both races of *D. pseudoobscura* could be found, and this student collected for him over the summer of 1932. Thus began a long experimental program on *D. pseudoobscura* (with Dobzhansky himself collecting samples for years). Historians have described how central this species was for Dobzhansky's pioneering work in evolution, for it led him to support the *biological species concept*, which defined species as a population consisting of a set of interbreeding subpopulations. Emphasis was placed on the population level; that is, he defined species by reference to the characteristics of the

2. For a discussion of the origin of this book, see Joe Cain, "Co-opting Colleagues: Appropriating Dobzhansky's 1936 Lectures at Columbia," *Journal of the History of Biology* 35, no. 2 (2002): 207–19. For Dobzhansky's life and career, see Mark Adams, ed., *The Evolution of Theodosius Dobzhansky* (Princeton, N.J.: Princeton University Press, 1994), and E. B. Ford, "Theodosius Grigorievich Dobzhansky," *Biographical Memoirs of Fellows of the Royal Society* 23 (1977): 58–89. William Provine, *Sewell Wright and Evolutionary Biology* (Chicago: University of Chicago Press, 1986), has an excellent discussion of Dobzhansky's writings on genetics.

3. Various *Drosophila* species were used in genetic experiments. See Robert Kohler, *Lords of the Fly: Drosophila Genetics and Experimental Life* (Chicago: University of Chicago Press, 1994).

population, rather than on a set of physical characteristics "typical" of an ideal individual. The definition also stressed the dynamic, rather than static, nature of species. The subpopulations are constantly in flux, but individuals from each subpopulation can successfully reproduce with others from the same or a different subpopulation.[4]

Due to a number of factors, however, one subpopulation can become separated and begin to develop *isolating mechanisms*. In time, the separation can become so great that it can lead to its complete reproductive isolation from the parent population; that is, individuals in the separated subpopulation can no longer successfully reproduce with individuals in the parent populations. When this happens, the subpopulation (or what is sometimes called a subspecies, variety, or race) has become a new species. This approach differed significantly from earlier ways of conceptualizing species by way of an averaged set of physical characteristics. Ultimately, Dobzhansky argued that because of their reproductive isolation, the two "races" described by Lancefield (which in gross anatomical traits looked identical) were actually *separate species*. It was a critical distinction that focused on the reproductive isolation among species, and the lack of such isolation among races. For the time, it was an unusual position, because to the naked eye, indeed even magnified, individuals from the two species looked identical.

Dobzhansky's position on race was quite radical. Race, according to him, is a "tool for description not of individuals, but of subdivisions of species."[5] Races differ in their gene frequencies and arrangements. A race, according to Dobzhansky, was not characterized by an overall physical description, or by a

4. There were several versions of the biological species concept. The most famous was formulated by Ernst Mayr. The modern synthesis and its followers used concepts like population to refer to various levels of generality. It has continued to be a vexing problem in biology, especially since groups of organisms are so very different. For our purposes, a species is a population of interbreeding individuals. Within the species, there are many local populations (subpopulations). When a local population is geographically distant and displays some differences in gene frequencies, it is called a subspecies, a variety, or a race. Theoretically, all the individuals of different races can interbreed, but there are cases where a species is spread over a large territory, and the races from the very farthest ends may have some reduced fertility. The point is that the races interbreed well with races near them, and therefore genes can travel the entire length of the range. In what follows, I have not made an effort to elaborate on the complex relationships of subpopulations, or populations within subpopulations. When I use the term *subpopulation*, I am referring to those populations of a species that are, or have been, geographically isolated or separated, and in which one finds some difference in some gene frequencies.

5. Theodosius Dobzhansky and Carl Epling, *Contributions to the Genetics, Taxonomy, and Ecology of Drosophila pseudoobscura and Its Relatives,* Carnegie Institution of Washington Publication 554 (Washington, D.C., Carnegie Institution, 1944), 138. Also see Theodosius Dobzhansky, "The Race Concept in Biology," *Scientific Monthly* 52, no. 2 (1941): 161.

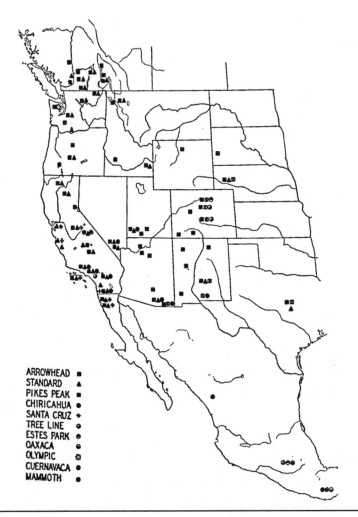

ARROWHEAD ■
STANDARD ▲
PIKES PEAK ▪
CHIRICAHUA ●
SANTA CRUZ ✦
TREE LINE ○
ESTES PARK ◉
OAXACA ◎
OLYMPIC ✺
CUERNAVACA ●
MAMMOTH ●

Dobzhansky

In 1937, Theodosius Dobzhansky published one of the first synthetic treatments of the modern theory of evolution. *Genetics and the Origin of Species* provided a genetic foundation for natural selection, the force Darwin claimed was responsible for the process of forming new species. Dobzhansky spent years studying the distribution of genes in populations of *Drosophila*, the common fruit fly. This map from Dobzhansky's book illustrates the geographical distribution of gene arrangements on one of the chromosomes of *Drosophila pseudoobscura*.

■ From Theodosius Dobzhansky, *Genetics and the Origin of Species* (New York, Columbia University Press, 1937), 94.

minor physical variation like color or size, but by the percentages of specific genes, or arrangements of genes. A race can include individuals with traits associated with other races; what matters are the *frequencies* of the genes that are responsible for the traits. By looking at an individual's characteristics, one might be able to determine the probability of its race, but only its probability. The individual could be from any race. Dobzhansky stressed that races are open systems and species are closed systems. He meant by this that since individuals of different species are not able to reproduce successfully, species do not regularly exchange genes. Individuals belong to one species or another.[6] In contrast, genes regularly flow from race to race since the individuals are interfertile. An individual might carry some genes that are frequently found in one race, but also carry genes that are most frequently found in a different race.[7] Since races contain many different traits, they can be defined in any number of different ways depending on what characteristics are of interest to the researcher.

These were significant distinctions, and when applied to humans, they suggested a new conception of the biological significance of race. Dobzhansky always quickly noted the inadequacy of our everyday conceptions of race, and the primitiveness of our knowledge of human genetics. The lack of consensus in anthropology over classification of human races reflected the difficulty of adequately defining them. Nonetheless, he believed that human populations had experienced sufficient geographical isolation to create populations with meaningfully different frequencies of certain genes. The vast amount of mixing maintained the status of races as populations of the same species. And, since genes can vary independently, by focusing on a particular gene (or expression), one can classify humans in quite a number of ways. Skin color doesn't correlate, for example, with blood type. A classification based on blood type would be different than one based on skin color, or eye color; hence, there was a certain arbitrariness in classifications. Because of the constant mixing, there can be no such thing as a "pure" race.

What, then, of the disharmonies that earlier geneticists had worried about? Dobzhansky held that races reflect a stage in evolution—they are subpopulations that, due to being geographically separated or to other biological pressures (adaptive pressure, genetic drift, etc.), have come to have discernibly different sets of gene frequencies than other subpopulations of the same species. The greater the differences in gene frequencies, and the more the subpopulation is geographically isolated, the greater the chance that the race will develop

6. Hybrids are either sterile or less fertile.
7. Dobzhansky and Epling, *Contributions*, 49.

complexes of genes adapted to a specific environment and begin to develop isolating mechanisms that restrict gene flow. If this goes on long enough, it may become a different species. But most of the time, normal interbreeding keeps all the races part of the same species; that is, the isolation hasn't become so severe as to restrict successful breeding. What about human races? Theoretically, races of humans might exist that, having been isolated for long periods, have developed *combinations* of genes with adaptive value. One can imagine that if individuals from different races that had these different *adaptive gene complexes* interbred, the result might be offspring with individual genes from each of the parents but not in the combinations that had given the individuals in the parental races some adaptive edge. These offspring would, presumably, be less fit than many of the individuals in the parental races. Dobzhansky argued that although gene complexes could favor survival and reproduction (i.e., be adaptive), and that hybridization could in theory lead to less well-adapted combinations, scientists had *no* empirical evidence of adaptive differences among human races. He described earlier fear of disharmonies as simply "farfetched."[8] Different human populations had mixed extensively. Dobzhansky used many examples to illustrate his point that this mixture had not resulted in disharmonious constitutions.

One might inquire why Dobzhansky had confidence that race mixing posed no serious biological problem whereas East and Davenport, who had established reputations in genetics, had stated otherwise. Here Dobzhansky's perspective, built from years of breeding *Drosophila*, mattered. A subpopulation (race) for Dobzhansky constituted an open system with substantial variation. He found that the frequencies of particular genes varied in subpopulations over seasons and over relatively small geographical locations. Mixing constantly occurred and was critical for keeping the species intact.

The process of evolution, of course, can lead to significant divergence, and with that, the creation of isolating mechanisms. Although Dobzhansky recognized many forms of isolating mechanisms, what they all had in common was that they caused crosses with individuals from different groups to have either no offspring or less fertile offspring. When it came to humans, *no* credible literature substantiated the claim that individuals from different populations exhibited any such difficulty in reproduction. There was speculation, to be sure: a common theme in nineteenth-century literature on race was the infertility or degenerate nature of the offspring of Caucasian and Negro crosses.[9] To Dob-

8. Theodosius Dobzhansky, "The Genetic Nature of Differences Among Men," in Stow Persons, ed., *Evolutionary Thought in America* (New York, George Braziller, 1956), 144.

9. A speculative literature existed that claimed mullatoes were infertile, or physically inferior,

zhansky, this literature was highly suspect, not only because of its crudeness and speculative nature, but also because it wasn't clear that earlier anthropologists (even with the help of geneticists) had the intellectual tools to identify adequately populations that constituted races. Dobzhansky, for example, did not believe the Jews constituted a race and thought that the term Negro in its American context was too hopelessly broad to be meaningful.[10]

At issue was the entire approach of defining different races, which rested on a methodology that the modern synthesis rejected. Physical anthropologists, zoologists, and early geneticists had categorized groups by tabulating the mean (average) values of a set of physical measurements. Davenport relied on an elaborate set of measurements that he used in comparing different populations.[11] He argued that the mean values of these measurements differed among races, and in describing the potential dangers of race mixing, he contrasted the mean sizes of different groups. Dobzhansky argued that comparing means had no value, for populations consisted of individuals, and mean values were the properties of groups. The variation and complexity of the possible combinations of genes, moreover, was so great that comparing the mean values of a few selected traits was a meaningless exercise. Humans shared so many genes, and the variation within a population was so large, that specific individuals from different populations might be more alike than individuals from the same population. Concentrating on mean values suggested some ideal type, a notion Dobzhansky and his colleagues, such as Ernst Mayr, were trying to drum out of existence in biology.

Dobzhansky's views on genetics and evolution undercut opposition to race mixing in several important ways. First, objecting to race mixing involved a category confusion. Race is a group property (different relative frequencies of some genes), not an individual property, so it is races that mix, not individuals. Second, races must mix; otherwise, they become separate species. Races of a species, including humans, have a normal degree of gene exchange with other races. All the races that exist today are the products of extensive mixing in the past. There are, consequently, no pure races. According to Dob-

---

to Negroes or Whites, but by the late thirties, this literature had no scientific credibility. An often cited text in this earlier literature on the problems of Negro/White "hybrids" is a translation of Paul Broca, *On the Phenomena of Hybridity in the Genus Homo* (London: Longman, Green, Longman and Roberts, 1864), which claims that the genital organs of the different races were of differing sizes and made certain crosses difficult or impossible (e.g., Negro male and White female).

10. See Dobzhansky, "Genetic Differences," 106–107, 155.

11. See Charles Davenport, *Guide to Physical Anthropometry and Anthroposcopy* (Cold Spring Harbor, Eugenics Research Association, 1927).

JOHN HERVEY ½N ½W   JENNIE THOMAS HERVEY ½N ½W

JOHN WHEELER ½N ½W

#3 MARGARET HERVEY WHEELER ½N ½W   #1   BROTHERS OF MARGARET H. WHEELER   #4   #2 SISTER OF MARGARET H. WHEELER

RUTH WHEELER ½N ½W

× MARJORIE WHEELER ¼N ¾W

JOHN HERVEY WHEELER ¼N ¾W

Variation within Races

Geneticists have often noted that there is more variation within a race than between races (even when we use the everyday concepts of race rather than more rigorous scientific ones). There are numerous ways to demonstrate this fact, but on a common sense level, it is easy to see when we consider the phenomenon of *passing*. Individuals of "Negro" heritage vary considerably in skin color, some being indistinguishable from many individuals who live in equatorial Africa, others being indistinguishable from many "Caucasians" living in the United States. Some of these latter have, over the years, established identities as "whites" and lived in white society without notice or comment. Novels, plays, and movies (e.g., Philip Roth's novel *The Human Stain*) have depicted the personal struggles of those who "pass" for white.

The children of mixed-race couples complicate the story even more. Skin color is determined by the combined action of several genes, operating in a complex fashion. Children of single race parents can display a range of variation, and the children of mixed-race parents can display an even wider range. This phenomenon was nicely documented by Caroline Bond Day, a student at Radcliffe College who worked with the famous Harvard anthropologist Earnest Hooton. She did a physiological and sociological study of mixed-race families, part of which she published in 1932. The images here are from that study and reflect the range of skin color within one family, from dark to light, as well as the range of facial features.

■ From Caroline Bond Day, *A Study of Some Negro-White Families in the United States* (Cambridge, Mass.: Peabody Museum of Harvard University, 1932), plate 16.

zhansky, even the most extensive attempt to separate human populations, the Indian caste system, shows no evidence of having produced genetically pure groups.[12] Third, when discussing race mixing, earlier naturalists, geneticists, and anthropologists had compared mean values of traits. This made no sense to Dobzhansky. Means are mathematical calculations and don't mix. Variation within any population is enormous, and looking at means obscures this wide set of differences. There are, for instance, "Negroes" who are whiter than "Caucasians." A fourth issue had to do with the empirical record. Whites mix with blacks, and there is no evidence of sterility in the offspring, or "disharmonious" constitutions resulting. Also, there is no evidence that the occasional poor combination of genes, mostly deleterious recessives, is more frequent in mixed marriages than in marriages within a race.

For Dobzhansky, then, earlier attempts to discourage race mixing based on alleged scientific arguments were invalid. Furthermore, his writings added to the shift in attitude that occurred in the 1940s because Dobzhansky was prominent in various symposia and edited volumes addressing the issue.[13]

Dobzhansky and Anthropology

Several of Dobzhansky's contemporaries who agreed with him wanted to abolish the use of the term *race* altogether in referring to human groups and replace it with the term *ethnic group*, as argued by Julian Huxley and Ashley Montagu. But major concepts in science, although often radically modified (e.g., inertia, heat, species, and nutrition), are not easily dropped, and it should come as no surprise that Dobzhansky did not see the value in eliminating a term that was central to his research. Although he accepted the view that current, everyday racial categories of humans were social constructions, and that races could be classified in many ways—the principal criterion being convenience—he nonetheless held that *races* had biological meaning. That is, humans group into Mendelian populations, groups of interbreeding individuals. These populations are mobile and in constant flux, but they are, nonetheless, real.

The argument was not limited to semantics. Those who preferred the term *ethnic group* considered such designations as purely social constructs. They were largely reacting to the bogus racial theories of the Nazis that depicted Jews as a separate race and divided Europeans into a set of different races. The issue, in many ways, had to do with what appeared to be confusion in the

12. Dobzhansky, "Genetic Differences," 105. This argument was repeated in many other articles in which he discussed human evolution.

13. See, for example, *Origin and Evolution of Man, Cold Spring Harbor Symposia on Quantitative Biology*, 23, 1950.

meanings of nationality, ethnicity, and race. As much as Dobzhansky abhorred the abuse of the term *race,* and as much as he realized the arbitrary nature of any particular classification of human races, he remained tied to his population biology perspective, which recognized species as consisting of many interbreeding subpopulations, traditionally called races. To agree that races were purely social constructs was to concede the field to those who misused science and, moreover, was not likely to accomplish any reduction in discrimination. For someone like Dobzhansky, who knew the dangers of letting social priorities dictate science, such a move was unacceptable and unwise. He was painfully aware that the research school of population genetics to which he belonged was under attack in the USSR (it would ultimately be abolished). Marxist attacks on Mendelian genetics demonstrated the damage that could be done to science when ideology overcame scientific evaluations.[14] Similarly, he was aware of how Nazi "racial science" had put social policy before rigorous science, and how scientific racism in the United States had deformed the life sciences for many years to support justifying slavery, limiting immigration, and outlawing race mixing. The answer, according to Dobzhansky, was not to give in to the misuse of science, but to educate the public in properly understanding scientific concepts.

Dobzhansky and Montagu sparred over the issue of substituting the term *ethnic group* for *race* for years. In a 1944 letter, Dobzhansky wrote,

> It is hardly necessary to say that I admire your book, and regard its function as contributing to sanity as important as any book can make itself to be. My criticisms, at least most of them, are due to not only academic fuss over details, but, and principally to fear of what use these details can be made of by the opposition. It is obvious I think that the racialists which were so overwhelmingly strong in USA before 1932 and before Hitler has made Madison Grant's view unpopular, have not disappeared since. They are just laying low and waiting for their day, which may come as soon as the war ends or even sooner. Just think what will then be written by the eugenists—every little slip of yours will be used to show that all that you wrote is wrong.
>
> It is for this reason that I can not subscribe to your campaign against the existence of races and for the ethnic groups or divisions or whatever you want to call them. The only way is to divest the word race of it [sic] emotional contents; and

14. See Zhores Medvedev, *The Rise and Fall of T. D. Lysenko* (New York: Columbia University Press, 1969); David Jaravsky, *The Lysenko Affair* (Cambridge: Harvard University Press, 1970); Valerii Soifer, *Lysenko and the Tragedy of Soviet Science* (New Brunswick, N.J.: Rutgers University Press, 1994); and Nils Roll-Hansen, *The Lysenko Effect: the Politics of Science* (New York: Humanities Books, 2005).

if we biologists can help in this, we shall justify our existence. Surely biologists by themselves can not do it 100%! The propagandist trick of making people swallow something under a different name might be useful in salesmanship and politics but I am afraid of its consequences in science.[15]

Montagu did not accept Dobzhansky's point of view and continued to argue against the use of *race* in anthropology. Although the two were personal friends and supported one another in many projects, they never came to agreement on the issue of human races. In 1961, Dobzhansky, somewhat lightly, wrote to Montagu about the latter's autobiography: "The chapter on 'Ethnic group and race' is, of course, deplorable. But let us say that it is good that in a democratic country any opinion, no matter how deplorable, can be published."[16]

The squabble with Montagu over terminology reflected Dobzhansky's concern with the social use of science. Among the architects of the modern synthesis in the United States, Dobzhansky was the most outspoken on social issues. He had collaborated with his Columbia University colleague L. C. Dunn in 1946 on a book written for the general public on the implications of modern genetics for social issues, and he continued to write for general audiences the rest of his life.[17] More important, Dobzhansky actively opposed those who tried to use evolution and genetics to argue against race mixing. His fears were justified, for some individuals attempted to enlist the modern evolutionary theory to oppose race mixing and support socially restrictive policies (such as segregation). For example, Dobzhansky reacted strongly to the publication in 1962 of Carleton Coon's *The Origin of Races*. Coon received his education at Harvard with Earnest Hooton, the leading physical anthropologist in the United States in the first half of the twentieth century. Hooton had written extensively on race and, using physical characteristics, sought to classify the different human races. Coon worked in that tradition and opposed the cultural anthropologists associated with Boas who thought of race in cultural terms. *The Origin of Races* attempted to synthesize the current information

15. May 22, 1944, Montagu Papers, American Philosophical Society. On Madison Grant, see chapter 2, footnote 11.

16. January 12, 1961, Montagu Papers. Dobzhansky was not alone in criticizing Montagu, and the Montagu Papers have responses from several zoologists and anthropologists who agreed with Dobzhansky. Montagu was not successful in popularizing *ethnic group* in anthropology, but, in a sense, he was on the winning side in that the use of the term *race* was increasingly seen as not having any biological meaning. See Nathaniel Gates, ed., *The Concept of "Race" in Natural and Social Science* (New York: Garland, 1997). The story is very complicated, and the issue is still quite controversial.

17. Dunn and Dobzhansky, *Heredity, Race, and Society* (New York: New American Library, 1946).

from the fossil record with the principles of evolution. It was a sophisticated treatment that discussed geographical variation and remained sensitive to the paucity of concrete evidence available.

The underlying theoretical thrust of the book, however, contained a controversial position: polycentrism, what we call today the multiregional origin of *Homo sapiens*. Coon drew on the writings of Franz Weidenreich, a German physical anthropologist who wrote extensively on the subject during the first half of the century. Weidenreich believed that early humans spread over a vast set of regions. Migration and mixing kept these populations from diverging into different species and kept the races of humankind continually in flux (i.e., there are no pure races associated with specific geographical regions). There were, nonetheless, some characteristics associated with particular regions, and this continuity was reflected in the fossil record. He rejected the idea that the different races of humankind represented separate evolutionary lineages (polygenesis) and stressed that humans had a single origin. Races are cross-fertile and interbreed in a normal fashion.[18] Weidenreich claimed that race mixing continues to occur naturally, and "in no cases have sexual aversions been manifested unless enforced by the interference of man himself."[19]

Coon took Weidenreich's theory in a rather different direction. Instead of envisioning a single human species spread out over a wide range that continued to develop in time but retained certain regional characteristics, Coon hypothesized that the ancestral *Homo erectus* had spread across the globe, had separated into five different subpopulations, and that these subpopulations each evolved independently into *Homo sapiens*. Moreover, Coon claimed that these five populations had crossed over into *Homo sapiens* at different times— Caucasoids early and Congoids later. In the introduction of his study, Coon made explicit the racist inference such a timetable implied. He stated that, "It is a fair inference that fossil men now extinct were less gifted than their descendants who have larger brains, that the subspecies which crossed the evolutionary threshold into the category of *Homo sapiens* the earliest have evolved the most, and that the obvious correlation between the length of time a subspecies has been in the *sapiens* state and the levels of civilization attained by some of its populations may be related phenomena."[20]

18. For a sympathetic and interesting discussion of Weidenreich, See Milford Wolpoff and Rachel Caspari, *Race and Human Evolution: A Fatal Attraction* (Boulder, Colo.: Westview Press, 1997). Weidenriech was forced to resign his position in Germany due to his Jewish background and came to the University of Chicago in 1934.

19. Weidenreich, *Apes, Giants and Man* (Chicago: University of Chicago Press, 1946), 2.

20. Coon, *The Origin of Races* (New York: Alfred Knopf, 1962), ix–x.

Like earlier students of race who constructed "scientific rankings," Coon opposed race mixing. In his concluding chapter, he claimed that when two races come into contact, one tends to dominate. Moreover, local populations resist the intrusion of outsiders, and social mechanisms come into being that encourage isolation of the populations. He cited Europeans in India, Indonesia, and Africa as examples. He then suggested that there is a genetic component of this separation: "Genes in a population are in equilibrium if the population is living a healthy life as a corporate entity. Racial intermixture can upset the genetic as well as the social equilibrium of a group, and so, newly introduced genes tend to disappear or be reduced to a minimum percentage unless they possess a selective advantage over their local counterparts."[21] Nature, then, keeps the races separated: "I am making these statements not for any political or social purpose but merely to show that, were it not for the mechanisms cited above, men would not be black, white, yellow, or brown. We would all be light khaki, for there has been enough gene flow . . . during the last half million years to have homogenized us all had that been the evolutionary scheme of things, and had it not been advantageous to each of the geographical races for it to retain, for the most part, the adaptive elements in its genetic *status quo*."[22]

Dobzhansky reacted strongly to Coon's monograph. He did so in part because it was the most up-to-date and complete review of the human fossil record and therefore a potentially influential book. The previous year, Dobzhansky had criticized a racist tract of Carlton Putnam, which had tried to support segregationist policies in the South by reference to anthropological literature that stressed the danger of race mixing.[23] Coon's book, Dobzhansky realized, could be used to bolster segregationist ideology. Although Dobzhansky was sympathetic to Weidenreich's theory, he disagreed with Coon's interpretation. In particular, he totally rejected the notion that five races of *Homo erectus* could have crossed over independently to become *Homo sapiens*. Dobzhansky argued that such a proposal made no sense from the standpoint of modern population genetics. A species is a closed system, and its subpopulations evolve together; once a population has evolved from an ancestor set of populations, another

21. Coon, *The Origin of Races*, 662.
22. Coon, *The Origin of Races*, 663.
23. See John Jackson, Jr., "'In Ways Unacademical': The Reception of Carleton S. Coon's *The Origin of Races*," *Journal of the History of Biology* 34 (2001): 247–85. Jackson explores the relationship of Coon and Putnam, most likely not known to Dobzhansky, and Dobzhansky's reaction to Coon. Also see Shipman, *The Evolution of Racism*, and Wolpoff and Caspari, *Race and Human Evolution*.

(separate) subpopulation could not conceivably follow the same route (especially in a different environment).

The controversy with Coon shows Dobzhansky's concern with the abuse of science for social agendas. Science had an important role to play in informing social debates with accurate information. Segregation in the South was often justified as an institution that inhibited race mixing, and the opposition to race mixing had been partly based on allegedly scientific grounds. Dobzhansky campaigned to eliminate the tie of racism to science and effectively used his work in evolutionary biology to support his position on race mixing.

5 The Modern Synthesis Meets Physical Anthropology and Legal Opinion

Dobzhansky's evolutionary writings had an important impact on physical anthropology, and this influence further undercut scientific opposition to race mixing. Along with the influence from cultural anthropology, these new ideas indirectly found expression in legal opinions that related to race. Montagu exemplified the anthropologists who used Dobzhansky's genetics to rethink basic concepts in physical anthropology.[1] Of equal importance was Sherwood Washburn, perhaps the leading figure in what came to be called the "new" physical anthropology that emerged on university campuses in the 1960s. In a 1951 classic article with that title, Washburn stated that physical anthropology had undergone a revolution due to the infusion of knowledge of evolution and population genetics. He characterized the "old" physical anthropology, as established in the United States by Earnest Hooton at Harvard and Aleš Hrdlička at the Smithsonian Institute, as "primarily a technique" that focused on measurement of the body, and on the correlation and classification of those measurements. Theory, in the old physical anthropology, consisted of a static attitude that sought to classify types.[2] In contrast, the new anthropology reflected knowledge of the modern synthesis and its application to the study of primate and human evolution. Race, for example, must be based on the study of populations, not on the averages of a set of individual measurements.

The New Physical Anthropology

Washburn represented a new movement in physical anthropology, one reflected in the pages of the *American Anthropologist* and the *American Journal of Physical Anthropology*, the leading journals in the field. He and Dobzhansky worked together, most notably by organizing symposia. For example, Washburn and

1. Like several of Boas's students, Montagu contributed to physical as well as cultural anthropology. He was more of a physical anthropologist than a cultural one.
2. S. L. Washburn, "The New Physical Anthropology," *Transactions of the New York Academy of Sciences*, series 2, 13, no. 7 (1951): 298–304, reprinted in Shirley Strum, Donald Lindburg, and David Hamburg, eds., *The New Physical Anthropology* (Upper Saddle River, N.J.: Prentice Hall, 1999), 1–5.

Dobzhansky organized the 1950 Cold Spring Harbor Symposia on Quantitative Biology, an annual and highly prestigious event that invited some of the leading life scientists to survey the state of various fields. What became volume 15 in this series brought together evolutionary biologists, geneticists, and anthropologists to discuss human evolution.

The symposium speakers repeatedly stressed the primitive state of knowledge in human genetics and the complexity of the problems. Some exciting work on the genetic distribution of blood groups was discussed at the meeting, but even this work revealed how much more there was to do. C. P. Oliver, the chairman of one of the sessions on the genetic analysis of racial traits, was blunt: "The gauntlet has been thrown to human geneticists. It is our privilege to determine more accurately the genetics of human traits, the more complex as well as the simple. If we can judge from the discussions of the past few days, the geneticists are aware of the scarcity of accurate data and the need to develop our knowledge about our traits so that a better analysis of human populations will be possible."[3] But there was a sense that the new biology could revitalize the study of race and put it on a firm footing. Joseph Birdsell, the famous anthropologist of Australian aborigines, optimistically claimed that population genetics could move physical anthropology beyond the flawed descriptive stage. He wrote, "Racial anthropology is not bankrupt, but it will require assistance in bridging the awkward gap between its descriptive phase of development and the new analytical phase lying ahead. The problems of human evolution and racial differentiation are essentially population problems, and their solution will be advanced by borrowing techniques of analysis from the vigorous field of population genetics."[4]

Ernst Mayr, one of the principal architects along with Dobzhansky of the modern synthesis, stated at the symposium that modern humans are comparatively homogeneous because of all the interbreeding that has occurred among different tribes and races. The result has been a large, interbreeding set of populations that are not only "connected everywhere by intermediate populations but even where strikingly distinct human populations have come in contact, such as European and Hottentots, or as Europeans and Australian aborigines, there have been no sign of biological isolating mechanisms, only social ones."[5] Opposition to race mixing, therefore, rested on social, not biological, objec-

3. C. P. Oliver, "Genetic Analysis of Racial Traits (III)," *Cold Spring Harbor Symposia on Quantitative Biology*, vol. 15, 1950, p. 245.

4. Joseph Birdsell, "Some Implications of the Genetical Concept of Race in Terms of Spatial Analysis," *Cold Spring Harbor Symposia on Quantitative Biology*, vol. 15, 1950, p. 259.

5. Ernst Mayr, "Taxonomic Categories in Fossil Hominids," *Cold Spring Harbor Symposia on Quantitative Biology*, vol. 15, 1950, p. 112.

tions, and anti-miscegenation supporters could no longer use anthropology to legitimate their position.

Ashley Montagu went further than Mayr at the symposium and stressed the evolutionary importance of race mixing. In the history of mankind, Montagu noted, populations of humans became isolated, and at the boundaries between different populations, intermediate populations emerged. These in turn underwent further isolation, and then hybridization, and then new isolation, just as in Dobzhansky's *Drosophila* experiments. The mixing that takes place between different populations produces new combinations that can be of great value for adaptation to the environment. So, the mixing results in possible characters or qualities that otherwise wouldn't appear in a population. Race mixing from this perspective not only is unobjectionable, but represents the source of evolutionary novelty and hence is considerably important.[6] Montagu thereby reversed by 180 degrees the opinions expressed by East in his *Inbreeding and Outbreeding*: that the crossing of individuals of different human races was to be avoided on biological grounds (OK for hogs and corn, just not people!).

By the sixties, Stanley Garn, in an article titled "The Newer Physical Anthropology," would write that anthropometry and typology were gone, replaced by the sort of research Washburn and others had been advocating.[7] Although this assessment may have been overly optimistic, it did reflect an important change and a related buttressing of the scientific argument against those earlier anthropologists, geneticists, anatomists, and biologists who had argued that race mixing was a dangerous experiment. Until the work of the modern synthesis, and its extension to the issue of race mixing, biological arguments concerning anti-miscegenation (for or against) had rested on a weak scientific base. Davenport's data were hopelessly inadequate. But equally, Boas had to admit that he had little direct evidence to show that race mixing did not pose a threat. The modern synthesis shifted the ground of the debate. The new *theory* changed the meanings of basic terms, particularly *species* and *race.* The question, "was race mixing deleterious or not," became conceptually incoherent given the new definition of species (as a set of intermixing subpopulations)

6. See M .F. Ashley Montagu, "A Consideration of the Concept of Race," *Cold Spring Harbor Symposia on Quantitative Biology,* vol. 15, 1950, pp. 315–34.

7. Stanley Garn, "The Newer Physical Anthropology," *American Anthropologist* 64 (1962): 917. The story is, of course, more complex. See Frank Spencer, "The Rise of Academic Physical Anthropology in the United States (1880–1980): A Historical Overview," *American Journal of Physical Anthropology* 51 (1981): 353–64; Frank Spencer, ed., *A History of American Physical Anthropology 1930–1980* (New York: Academic Press, 1982); and Strum, Lindburg, and Hamburg, *The New Physical Anthropology.*

and race (a subpopulation of a species). Mixing among the different subpopulations of a species was the *normal* course in nature. Different subpopulations did not represent different types; indeed, extensive variation was the norm for populations and subpopulations. Montagu had used this position as early as the forties, and it gradually came to be accepted by anthropologists.

By the mid-1960s, modern synthesis dominated biological thought in the United States, and population genetics became increasingly incorporated into the general university curriculum. The new biological perspective provided the scientific background through which issues of race had to be considered. Courts, in cases leading up to the landmark *Loving v. Virginia*, referred to the biological "facts" as important. The shift of the scientific community's attitude, therefore, was critical in reorienting the basic assumptions that had provided justification for anti-miscegenation laws.

Loving v. Virginia and Previous Legal Opinions

As historians have noted, the legal changes that led to *Loving v. Virginia* reflected many shifts in American culture: a growing respect for blacks, a liberalization of personal mores, and repugnance to Nazi racial laws. Physical anthropology, the new theory of evolution, and genetics at the same time shifted the scientific understanding of race in a way that undercut earlier arguments supporting anti–race-mixing laws. Since earlier legislation had sometimes cited biological reasons to justify anti-miscegenation laws, the new science held potential legal significance. The courts ultimately caught up with the culture and science, both in overturning state anti-miscegenation laws as unconstitutional and in demonstrating to various state legislatures the untenable nature of their laws and inspiring their repeal.

The California Supreme Court was the first court in the twentieth century to strike down an anti-miscegenation law. In the landmark 1948 case of *Perez v. Sharp*, the court ruled that California's anti-miscegenation law violated the United States Constitution.[8] Although the vote was close—4 to 3—Justice Roger Traynor's majority opinion reflected a new and bold approach to the issue of race mixing. The case involved Andrea Perez, a white woman, and Sylvester Davis, a Negro man, who were denied a marriage license on the basis of the existing California law prohibiting marriages between white persons and Negroes or mulattoes (Civil Code, section 69). Traynor's opinion rested on his reading of the due process clause of the Fourteenth Amendment to include

---

8. *Perez v. Sharp*, 32 Cal. 2d 711 (Cal. 1948).

the position that marriage is a fundamental right. "There can be no prohibition of marriage except for an important social objective and by reasonable means."[9] Moreover, "Race restrictions must be viewed with great suspicion,"[10] for the Fourteenth Amendment was adopted to prevent state legislation designed to discriminate on the basis of race.[11] In his ruling, Traynor explored the assumption that supported the anti-miscegenation laws of California and other states: that the mixture of Caucasian with non-Caucasian would lead to inferior offspring, physically and mentally. He rejected the assumption by referring to "modern experts" who repudiated Mjoen and Davenport's contention that race mixing produced disharmonious crosses,[12] and by referring to Herskovits, Huxley, Myrdal, Garth, Klineberg, and Benedict, who argued that there is no scientific basis for the commonly held view that one race is superior to another. Traynor's opinion dismissed the list of physical maladies submitted by the defense for the Los Angeles County Clerk as evidence of the inferiority of Negroes by noting that "most, if not all, of the ailments to which he refers are attributable largely to environmental factors."[13] He further added that "generalizations based on race are untrustworthy in view of the great variations among members of the same race."[14] Although Traynor did not specifically refer to the modern synthesis, his opinion reflected the new genetics that informed his sources and rejected the earlier scientific literature that had legitimated anti–race-mixing sentiment.

Traynor cited other reasons for rejecting California's anti-miscegenation statute, for example, that the statute was "too vague and uncertain." The law declared void "all marriages of white persons with Negroes, Mongolians, members of the Malay race or mulattoes."[15] Traynor cited Boas's contention that the number of races varies depending on what system of classification one uses: from three or four to thirty-four. The legislature's classification was based on the system proposed by Blumenbach in the eighteenth century. Moreover,

9. *Perez,* 32 Cal. 2d at 714.

10. *Perez,* 32 Cal. 2d at 718.

11. The Fourteenth Amendment grew out of a concern for the individual rights of blacks in the South after the Civil War. States had enacted legislation that excluded blacks (serving on juries, voting, etc.), and the Fourteenth Amendment protected individual rights from state encroachment. At first interpreted narrowly, the Fourteenth Amendment and its equal protection clause came to be interpreted broadly in the 1950s, and is a foundation for an important part of civil rights law.

12. See chapter 2, footnotes 14 and 28; chapter 3, footnote 3.

13. *Perez,* 32 Cal. 2d at 722.

14. *Perez,* 32 Cal. 2d at 722, 723.

15. *Perez,* 32 Cal. 2d at 728, 729.

the legislature had made no provision for how to apply the law to those of mixed-race ancestry. In sum, he held that the law was not only too vague and uncertain, but also that it violated the equal protection of the law clause of the Fourteenth Amendment of the U.S. Constitution. Although a dissenting opinion strongly supported the anti-miscegenation law based on the eugenic and racist views popular earlier in the century, the state did not appeal the ruling.

It would be almost twenty years before the United States Supreme Court ruled that all state anti-miscegenation laws were unconstitutional, but fourteen states repealed their statutes between *Perez v. Sharp* and *Loving v. Virginia*. The first to do so, Oregon, was influenced by the *Perez v. Sharp* decision.[16] The new population genetics and physical anthropology were not central in all the written judgments. But given how the earlier anti-miscegenation laws had been legitimized by reference to the scientific racism of Davenport and others, the new science played an important, if sometimes off-stage, role. Science could no longer be used to argue credibly for the biological importance of preventing race mixing.

Loving v. Virginia represents a terminus in the legal restrictions on race mixing in the United States. The Supreme Court's ruling did not come as a surprise. This was the Earl Warren court, whose *Brown v. Board of Education* in 1954 marked an important turning point in antidiscrimination in U.S. history. Although *Brown v. Board of Education*, which ruled that separate schools for black students was a violation of the Constitution, did not treat race-mixing or anti-miscegenation laws, the issue clearly resided in the background. Keeping white and black children separate represented the principal reason for segregated schools—separate to avoid, or reduce, race mixing. The court used social science to establish the deleterious psychological effects of segregation on black children. Notably missing from the court discussion, however, was the scientific racism that would have informed the court had this case been pursued in the 1920s or 1930s.

The decision in *Loving v. Virginia* also came just three years after *McLaughlin v. Florida*, where the court struck down a Florida statute (Sec. 798.05) that stipulated: "Any Negro man and White woman, or any White man and Negro woman, who are not married to each other, who shall habitually live in and occupy in the nighttime the same room shall each be punished by imprison-

16. See Matthew Aeldun Charles Smith, "Wedding Bands and Marriage Bans: A History of Oregon's Racial Intermarriage Statutes and the Impact on Indian Interracial Nuptials" (master's thesis, Portland State University, 1997), 154–55.

ment not exceeding twelve months, or by fine not exceeding five hundred dollars."[17] The equal protection clause of the Fourteenth Amendment was cited as basis for overturning the statute. Although the court sidestepped the issue of Florida's more general anti-miscegenation law because the statute did not deal specifically with interracial marriage, the court noted, "The central purpose of the Fourteenth Amendment was to eliminate racial discrimination emanating from official sources in the states. This strong policy renders racial classifications 'constitutionally suspect' . . . and subject to the 'most rigid scrutiny' . . . and 'in most circumstances irrelevant' to any constitutionally acceptable legislative purpose."[18] The decision then noted a set of cases in which racial classification had been judged invalid: voting and property records, designation of race on ballots, segregation in public parks, and segregation in the public schools.

After *McLaughlin v. Florida*, the popular press focused considerable attention on the issue of interracial marriage, and various articles speculated that the Supreme Court would soon judge the subject. A 1967 article in the *Atlantic Monthly*, for example, noted that although the California Supreme Court had overturned its anti-miscegenation law, nineteen states still had them on the books. It went on to state that "no other civilized country has such laws except the Union of South Africa."[19] Pointing out the absurdity of the Virginia law, the author went on to ask, "Is it not alarming to know that in 1965 the new U.S. congresswoman from Hawaii, who is of Japanese descent, and her Caucasian husband could be criminally prosecuted under Virginia law if they were to reside there while Congress is in session?"[20]

Chief Justice Warren delivered the opinion of the Supreme Court on June 12, 1967. The case concerned Mildred Jeter, a black woman, and Richard Loving, a white man, who together had crossed into Washington, D.C., to be married in June 1958, and then shortly afterward returned home to Virginia, where they continued to live. They were subsequently arrested and charged with violating the Virginia ban on interracial marriage. They pleaded guilty, and the trial judge suspended their one-year jail sentence on the condition that they leave the state and not return together for a period of twenty-five years. The judge stated in his opinion that "Almighty God created the races white,

17. *McLaughlin v. Florida*, 379 U.S. 184 (U.S. 1964)

18. *McLaughlin*, 379 U.S. at 192.

19. William D. Zabel, "Interracial Marriage and the Law," *Atlantic Monthly* 216, no. 4 (1965): 75.

20. Zabel, "Interracial Marriage and the Law," 77.

Loving v. Virginia

The U.S. Supreme Court decision in *Loving v. Virginia* marked a landmark legal case. Shown here are Mildred Jeter (1939–2008), a black woman, and Richard Loving (1933–1975), a white man, who lived together in Virginia and were married in the District of Columbia in 1958. Their legal battles, supported by the ACLU, resulted in the famous Supreme Court ruling of 1967. The decision has come to stand as a symbol of progress in attitudes toward race mixing in the twentieth century.

Across the country each year, people celebrate June 12 as Loving Day (www. lovingday.org), which organizers hope will become a nationally recognized day. In 2008 there were sponsored events in New York, Los Angeles, Chicago, Washington, D.C., Seattle, Portland (OR), Eugene, San Francisco, and Memphis, plus, of course, many private celebrations throughout the country. Loving Day has its own page on Facebook, MySpace, and YouTube.

■ Mildred Jeter and Richard Loving, © Bettmann/CORBIS.

black, yellow, malay and red, and he placed them on separate continents. And but for the interference with his arrangement there would be no cause for such marriages. The fact that he separated the races shows that he did not intend for the races to mix."[21] Due to the new genetics and evolutionary biology, the old scientific racism no longer provided a respectable base from which to object to interracial marriage; therefore, the only remaining appeal was to a higher authority, one beyond empirical verification.

21. *Loving v. Virginia*, 388 U.S. 1 (U.S. 1967), at 3.

The Lovings moved to Washington, D.C., and began a set of appeals and legal moves that resulted in the U.S. Supreme Court's hearing their appeal. They had been convicted under two statutes in Virginia, one prohibiting and punishing a white and black person who go out of state to be married with the intention of returning, and the second prohibiting and punishing inter-marrying between a white and a black (a felony punishable by a minimum of one year and a maximum of five years in jail). The Court ruled that such statutes were based solely on racial distinctions, and noted that the Court had consistently repudiated such distinctions as odious to a free people. Referring to *McLaughlin v. Florida*, Earl Warren noted that two members of the Court (Potter Stewart and William O. Douglas) had previously stated that they could not conceive of a valid legislative purpose that used a person's skin color to determine a criminal offense. Warren concluded, "There can be no doubt that restricting the freedom to marry solely because of racial classifications violates the central meaning of the Equal Protection Clause."[22] Marriage, according to Earl Warren, is one of the basic civil rights of man, and the "Fourteenth Amendment requires that the freedom of choice to marry not be restricted by invidious racial discriminations. Under our Constitution, the freedom to marry, or not marry, a person of another race resides with the individual and cannot be infringed by the State."[23]

Loving v. Virginia made the enforcement of state anti-miscegenation laws and state constitutional provisions impossible, and the remaining fifteen states slowly repealed them (Alabama was the last, in 2000).

The dramatic shift in the sixties of legal opinion on race mixing was not lost on the generation of students who were coming of age in the new civil rights era. Although the decision in *Loving v. Virginia* wasn't until 1967, many students in the early sixties had already concluded that there was no valid legal basis for anti-miscegenation, and that efforts by the establishment, be they legislatures or college administrations, to frustrate interracial mixing reflected outdated perspectives and outdated knowledge.

22. *Loving*, 388 U.S. at 12.
23. *Loving*, 388 U.S. at 12.

6 University Campuses in the 1960s

While the courts were moving toward *Loving v. Virginia*, faculty and students on many college campuses across the United States had already altered their views on interracial relationships. By the time of the landmark Supreme Court decision, the civil rights movement was well under way, and university courses in genetics and anthropology undercut whatever vestiges remained of scientific racism. Race had come to be discussed in many classes as a socially constructed term, not a category that reflected some biological reality. The emergence of mixed-race couples on campuses belonged to a broader breakdown of taboos. For liberal students, opposition to interracial dating suggested a carryover of an antiquated system of mores. A look at a few campuses will reveal student tension over interracial dating and university policy.

Before the sixties, African Americans who attended college did so primarily at historically Black colleges and universities, but by the close of the decade, they were enrolled in institutions that were predominantly white. State laws, like Pennsylvania's Fair-Education Opportunities Act of 1961, eliminated quotas restricting the number of applicants according to racial, national, or religious origins. At the same time, the number of African Americans continuing on for higher education increased and in the following decade climbed even higher. As one might expect with such change, new social issues emerged as a serious part of the American educational landscape. With increased contact between members of different races on campuses and in classes, interracial dating and marriage became inevitable. The Reverend Farley W. Wheelwright (approvingly) told his Long Island congregation on September 16, 1963, that when schools (along with restaurants and hospitals) were integrated, miscegenation would be likely and be accepted.[1] He may have been thinking of the recent story that ran in the *New York Times* about Charlayne Alberta Hunter, the first "Negro" girl to enter and graduate from the University of Georgia, who had married a white southern student she had met while they were in school

1. "Unitarian Cleric Who Visited Georgia Backs Miscegenation," *New York Times*, September 17, 1963, 63.

together. To be sure, that same year former president Harry Truman (who had supported integration) stated publicly that he did not believe whites and blacks should marry and that intermarriage ran counter to scripture. But the 1960s civil rights movement had sensitized people to the abuses of segregation and the mindset that had supported it. Scientists had undercut previous scientific efforts legitimating anti-miscegenation sentiment, and many church groups condemned the existing laws forbidding interracial marriage. The National Catholic Conference for Interracial Justice issued a strong statement in November 1963 arguing that "Interracial marriage is completely compatible with the doctrine and canon law of Roman Catholicism . . . Races do not marry. Nations do not marry. Classes do not marry. Only persons marry." Moreover, any family or state efforts to abridge this right "should be condemned."[2] Some southern chapters of the Knights of Columbus were singled out for condemnation. Two years later, the General Assembly of the United Presbyterian Church also declared that there were no scriptural or theological grounds for prohibiting interracial marriage.

But after so many years of strong opposition, and with racism a real and powerful prejudice, interracial couples still faced formidable problems in the sixties. Just finding housing, even in a large northern city, could pose a significant challenge. The *New York Times* ran an extensive article in October 1963, detailing the increase of interracial couples found in the city and describing the discrimination and obstacles those couples faced.[3] Even in ultraliberal enclaves, such as Greenwich Village, mixed-race couples expected to be met with stares, if not more overt forms of disapproval. Social mores change slowly.

Reactions on Campus: Syracuse, Ohio State, Indiana, Berkeley

Campuses were often at the forefront of change and consequently at the center of controversy. Let's consider a few large schools that had several black students: on the East Coast, the private Syracuse University and the semipublic University of Pittsburgh; in the Midwest, the urban Ohio State University and the rural Indiana University; and on the West Coast, the often trend-setting University of California, Berkeley. These are just a sampling of institutions, but they broadly represent the college scene in the sixties.

Syracuse University had a growing African American student population and a significant foreign student enrollment from Africa and Asia. Some of these students had difficulty finding housing in the city. Housing discrimination came increasingly under fire. As one student wrote in the student news-

2. "Catholics Uphold Biracial Couples," *New York Times,* November 18, 1963, 47.
3. "Negro-White Marriages in Rise Here," *New York Times,* October 18, 1963, 2, 18.

paper, the *Syracuse Daily Orange*, "Let not Syracuse be complacent and sympathetic about the South when graduate foreign students are refused housing
because of their color."[4] University administrators tried to resolve off-campus
problems as best they could. They also stumbled on a problem closer to home.
In the spring of 1961, roughly three thousand students staged a demonstration
to protest reports that the university was interfering with mixed-race dating.
Articles in the *New York Times* quoted a university official who thought the
demonstration was "a typical spring outbreak and would probably end soon."[5]
The situation, however, was not just spring exuberance. The dean of women
believed it was her responsibility to monitor the dating practices of coeds.
While the dean claimed that she valued the "rich opportunity on a campus like
ours which represents so many races, creeds, and nationalities for students to
get acquainted with people who represent varying backgrounds," she also felt
it her duty to keep parents informed "whenever it seems advisable that parents
know more about the dating habits of their daughters."[6] That was a not-so-
cryptic code for saying that parents needed to be informed if a girl went out
more than once with a student of a different race. Not all students shared the
view that parents needed to be informed. One wrote the *Daily Orange* that
she felt it "was none of their business."[7] Students were beginning to move
away from an easy acceptance of the *in loco parentis* philosophy of university
administration, and they were also being sensitized to the national mood over
civil rights. As an editorial in the student paper put it, the issue was one of
the university "overstepping and abusing its powers . . . Moreover, if mama
is afraid to have her girl dating somebody from across the street she should
keep her home or send her to the University of Mississippi where they can
really protect her."[8] (This was a year before James Meredith forced the issue of
integration at the University of Mississippi.) Students at Syracuse were quick
to castigate the administration for treating mixed-race dating in a manner that
seemed to be in keeping with the discrimination practiced in the South. As
another letter to the editor put it, "What could be more hypocritical? What
could more contradict the lofty aims extolled by chancellor Tolley? How can
we hope to develop as mature and free-thinking individuals when this very
development is hampered by the standards of society? We cannot condemn

4. "White Supremacy: A Losing Battle," *Syracuse Daily Orange*, November 16, 1960, 2.
5. "Syracuse Students Continue Protests," *New York Times*, May 7, 1961, 86.
6. "Policy Inhibiting Mixed Dating Disclosed to DO," *Syracuse Daily Orange*, May 5, 1961, 1.
7. "Policy Inhibiting Mixed Dating," 1.
8. "Dating Policy," *Syracuse Daily Orange,* May 5, 1961, 2.

the South for its narrow-minded views on segregation when we ourselves do not fully accept this ideal."[9]

The upper administration moved quickly to diffuse the situation at Syracuse. The vice president denied that there was any "official university policy discouraging dating between students of different religions and races."[10] The dean of women, however, initially stayed unrepentant and stuck by her original statement. Ultimately, a group of administrators and students came up with the idea of creating a grievance board to hear and investigate student complaints. At this point, the dean retreated somewhat from her earlier position and stated that, in fact, no actual letters had been sent to parents about mixed-race dating. Moreover, she promised to look into any discriminatory practices by head residents in dorms.

But there was a long way to go. In the fall, the student newspaper ran an editorial advocating a change in the Greek system. The paper noted that only six black students were members of the thirty fraternities (and all six were in nominally Jewish fraternities). Like other universities across the country, old prejudices were beginning to be unacceptable. By December, students were taking part in Freedom Rides to the South.

At Ohio State, in Columbus, a city that had a significant black population, university administrators worked with students to combat racism. During the summer of 1960, the university dropped from their list of off-campus housing the names of rentals that discriminated, and the entry form for the university removed any reference to race or religion. Early in 1961, the university required fraternities to remove restrictive clauses or face loss of university recognition (by the following year all had complied).

But the environment was still charged. Local black leaders warned in 1965 that Columbus was ripe for racial outbursts, and that summer an African student who was walking with a white girl on one of the city's streets was attacked by some locals and beaten unconscious. One of the attackers was reported to have yelled at the girl, "It's high time you realize that you shouldn't be out with a Negro."[11]

The dean of women at Ohio State strongly supported the *in loco parentis* policy, and it wasn't until 1966 that coeds won the right to a relaxed dress code and a less restrictive evening curfew. Perhaps because the administration had been proactive in its antidiscrimination activities, there did not appear to be a

9. "University Dating Regulations Hit," *Syracuse Daily Orange*, May 5, 1961, 2.
10. "University Denies Control of Dating," *Syracuse Daily Orange*, May 8, 1961, 1.
11. "White Men Beat African Student," *Ohio State Lantern*, August 12, 1965, 1.

public tension about interracial dating, at least not in any manner that caused it to become a major campus issue. Race, however, remained a highly charged topic, and interracial dating was one of many issues that bothered students.

Even at schools like Indiana University, which were in small, rural, predominantly white towns, racial issues surfaced, often over housing and social interaction between the general student body and foreign students, particularly those from Africa, or black athletes. In the early 1960s, for example, letters to the editor in the *Indiana Daily Student* discussed housing discrimination in Bloomington. More telling of the racial attitudes was a letter concerning a new application form for dorm space. The new form no longer asked if a student would be willing to room with a person of a different race, and some students were concerned that they might be placed with a black roommate. One letter rejected this "attempt at integration," conceding that living next to a black student might be acceptable, but not with one.[12]

University administrations sometimes inflamed racial issues on campus, but on some campuses, they diffused potential problems. This is well exemplified on the West Coast at Berkeley, where the student body included a wide range of social attitudes. The campus was a highly political one, with students supporting overt Communist organizations as well as the American Nazi Party. Letters to the student newspaper, the *Daily Californian*, reflected this range of ideas, and some of the letters expressed opposition to race mixing. In one letter from 1962, a sophomore wrote that prejudice against Negroes was a problem that needed to be solved on an individual level, but "This is to say nothing about the 'sin' of intermarriage, a situation which both races frown upon in disgust."[13] The university administration actively fought discrimination and acted quickly in cases that were public. For example, in 1963 the Berkeley Jaycees created a furor when it asked Lynn Mark Sims, the junior class president, who was black, to remove himself as an escort for one of the contestants in the queen competition for the 18th Berkeley Football Festival. The sponsors of this event invited coeds from across the country, and some of the entries were from white southern schools. Some individual in the Jaycees' pageant decided that given the "sensitivity" of race relations at the moment, it would be best if Sims not escort one of the contestants (who might be from the South). Afterward, there was a lot of finger pointing, and it is not clear who was actually to blame. Sims did not make a fuss; he just left and attended the game. The *Daily Californian* ran a story describing the incident, and quickly after, there was a considerable outcry. The dean of men moved swiftly. He issued a statement

12. Letters to the editor, *Indiana Daily Student*, October 12, 1963, 4.
13. Letter to the editor, *Daily Californian*, April 11, 1962, 8.

that the action went totally against university antidiscrimination policies, and that if the Jaycees did not guarantee that it wouldn't happen again, the student group would withdraw from participation in the festival. (The student group that provided the escorts quickly issued a statement affirming their agreement with the dean's statement but noting that the decision to withdraw from the festival would have to be made by the student group, not a university administrator. This was Berkeley, after all.) The Jaycees ultimately apologized and promised it would not happen again, but the issue continued to be discussed on campus. One alumna wrote that equal opportunity on campus did not "yet mean providing white girls as 'dates' for Negro men."[14] Her suggestion was that next year, the festival should invite a Negro queen candidate to be escorted by one of Berkeley's "fine Negro students."[15] Subsequent letters reacted to the "bigotry" of the suggestion and its "fear of miscegenation."[16] As it turned out, the Jaycees discontinued the annual football festival, which had been declining in popularity anyway. They did not refer to the controversy in their cancellation announcement, but they had received a tremendous amount of negative attention, and it is difficult to imagine that this was not an important factor in their decision.

What set Berkeley apart from many other campuses was the enlightened and swift action of its university administrators. Or at least some of them. The dean of students (formerly dean of women) had not played any public role in the issue. She was a strong supporter of an *in loco parentis* policy, which continued to be manifest in areas like curfew hours for women. She retired from Berkeley in 1965 during the general unrest in American universities and during the "free speech movement" controversy at Berkeley, a dispute between students and administration that in serious ways marked the beginning of the student activism of the sixties.

University of Pittsburgh

What about the semipublic University of Pittsburgh, where this story began? Student concern about the administration's attitude toward interracial dating belonged to a set of related issues centered on the *in loco parentis* role of administrators and what some perceived to be a form of institutionalized racism. Not surprising, the dean of women's office was the flash point. Curfews for women, dress codes (again, for women), prohibition of any visit by a coed to the men's dorms, and similar rules seemed antiquated toward the mid-sixties,

14. Letter to the editor, *Daily Californian*, September 25, 1963, 9.
15. Letter to the editor, *Daily Californian*, September 25, 1963, 9.
16. Letter to the editor, *Daily Californian*, September 30, 1963, 9.

but the dean of women still defended them as part of her responsibility as a "substitute parent."

But it was just this *in loco parentis* role that students in the sixties contested. Numerous articles in the student newspaper reflect that concern. For example, a *Pitt News* editorial in the fall of 1963 that covered Dean Rush's appearance at a Student Union Forum Series Administrative Coffee Hour, cautioned students to phrase their questions in a positive manner. That is, do not merely fire hostile questions, like why do girls have curfews, but ask something like, "Could not a graduate system of curfews be established which would be in keeping with the growing maturity of the students?" The editorial wanted to keep the discussion on a higher plane than just personal attack, and it noted that in previous editorials, the paper had not labeled the dean an ogre but had "merely expressed our disapproval of the *in loco parentis* (substitute parent) theory which governs her actions."[17] This general rejection of the administration as a substitute parent was at the heart of many student complaints, from the restriction of all unmarried female students to living in the dorms to the "interference" in interracial dating.

Although not a common occurrence, since mixed couples were still unusual, interracial dating was, nonetheless, a hot button issue. In October 1962, for example, the student government considered sending a letter to the University of Mississippi expressing disapproval over the violence that attended the forced admission of James Meredith to the university (which had required U.S. Marshals to enforce). An editorial in the *Pitt News*, "Our Own Back Yard," questioned the right of Pitt students "to cast stones . . . in view of certain facts concerning our own campus."[18] The editorial referred to (among other things) discrimination against blacks in their attempts to rent apartments in the areas near the university, resistance of some white students (or their parents) to being assigned black roommates, and an incident that "blemishes the University's recent past." The incident, which had "been developing for more than a full trimester and has long been hidden in veils of rumor," involved an "African Negro" who had been dating a white girl. According to the foreign student, when the "administrators governing the social regulations of the co-ed learned of the interracial relationship, they promptly summoned her to their office." She was then allegedly sent to a school psychiatrist, "but it was found that she was suffering from no emotional problems arising" from the relationship. The foreign student also claimed that "when the girl's parents arrived on campus to see her honored at the Spring Convocation, they were invited to

17. Editorial, *Pitt News*, November 6, 1963, 4.
18. "Our Own Back Yard," *Pitt News*, October 10, 1962, 4.

Dean Helen Pool Rush

Dean Helen Pool Rush, shown here, considered her *in loco parentis* role central to her mission as dean. She did not think she was restrictive, but thought that she, and the policies of her office, reflected the mores of the time. In an interview printed in the *Pittsburgh Press* after her retirement, Dean Rush stated, "I can look back and laugh at some of the things we did . . . I think how could we have been so stupid? But it wasn't stupid in terms of the life then . . . I'm sure we never thought of ourselves as restrictive." As a quaint example, she mentioned how undergraduates were instructed not to "express affection" while on the cathedral lawn, to avoid upsetting riders on the street cars that might be passing by!

■ Helen Pool Rush, dean of women (1942–1965). Courtesy of the Archives Service Center, University of Pittsburgh.

the administrators' office to discuss the 'situation.' Further correspondence between the parents and the University resulted in the transfer of the co-ed to a school in her home town."[19] The story circulated in the wider Pittsburgh community, and the black newspaper ran the story, "Test Co-ed's Sanity After Dating Negro," on its front page.[20] (The same issue of the *Pittsburgh Courier* carried stories about James Meredith and John F. Kennedy's speech supporting desegregation of the University of Mississippi.)

In a letter to the *Pitt News*, the president and vice president of the student government discussed the case. They wrote, "As a result of looking into the matter, we can quite honestly state to the student body that the report pre-

19. "Our Own Back Yard," 4.
20. *Pittsburgh Courier*, October 20, 1962, 1, 4.

sented in *The Pitt News* was a compilation of half truths, unjustified insinua-
tions, and actual falsity." Unfortunately, they continued, "the relevant factors
are of a very personal nature to the individuals involved and to delve into the
matter merely to satisfy curiosity or to prove a point would be an unforgivable
breach of privacy."[21] In response, a letter was sent to the student newspaper by
two students who claimed to know more of the story, and who felt the student
government leaders were mistaken. The letter claimed that

> A policy of interference with the personal relationships of "Pitt Women" and for-
> eign students is a little-discussed but very important part of the "social regulations"
> carried out by the Dean of Women. We first saw evidence of this fact when a young
> woman of our acquaintance was told by her head resident that she shouldn't eat
> in the cafeteria with the African students because "it did not look good." Soon
> after their campaign against the "white girl student" in question began. They tele-
> phoned the roommate of the girl and told her to drop hints to the "white girl stu-
> dent" that she ought to go to the school psychiatrist. Later, because the roommate
> refused to be a party to such action, the suggestion was made more directly. What
> followed was related precisely in the editorial. The actions taken by the "African"
> student after the removal of the girl from school are as follows, in his own words:
> "I've been like a little dragon going around to all the deans' offices. Everyone was
> scared to discuss this thing with me. When I came in, they turned their heads and
> stared at the tops of their desks while they talked with me. There's a black cloud
> hanging over the University, and everybody's scared of it."

The letter goes on to state that the writer had discussed the student government leaders' letter with one of its authors and that the person had not spoken
to "the girl, or even to the Dean of Women. She only talked to Asst Chancellor
Rankin."[22]

Alan Rankin, the assistant chancellor for student and general affairs, felt
the allegations of discrimination by the university were unfair. In a memo to
Litchfield, he wrote, "There *have been* individual acts of discrimination against
this African student (his door was painted red after he started dating the white
girl—a sign of death in Africa). It has been very hard for him to sort out individual attitudes and distinguish them from official attitudes. And apparently,
because the human animal is often a contrary being, there are students on the
campus who enjoy believing the worst about other human beings. Since we are

21. "S G Leaders Take Stand on Editorial," *Pitt News*, October 17, 1962, 4.
22. "Back Yard Hypocrisy Criticised by Students," *Pitt News*, October 29, 1962, 4.

not at liberty to explain publicly some things we know about the girl involved, it becomes especially frustrating."[23]

The story may have been more complicated than either the *Pitt News* or the *Pittsburgh Courier* knew, but nonetheless, it continued to circulate for years. Although there may have been a side of it that mitigated the rumor, the fact that it had such long legs suggests that what students experienced in their dealings with the Dean of Women's Office led them to accept the story as fact.

Rankin's memo also raises the distinction between individual acts of discrimination and institutional discrimination. Rankin (and Litchfield) took what they considered to be significant measures to fight discrimination at the university and in the surrounding city. They set up a committee against discrimination at the university, and they worked with the city's Commission on Human Rights to pressure landlords who discriminated in renting and with the Pennsylvania Human Relations Commission to protest cases of blacks not being served in eating establishments in the state.

But part of Rankin's job was to put the best face on situations, and even though he and Litchfield should be credited with furthering the fight against discrimination, the record is not altogether unambiguous. In an interview with the *Pitt News*, reported in October 28, 1963, Rankin discussed a new "administrative appeals committee to hear cases of alleged discrimination from members of the University community." Rankin was quoted as saying, "Our personal view is that there is no discrimination on campus. Officially it's contrary to University policy, and I don't think it exists on a personal level either. I don't think the committee will have much business. I hope not. This will be one way to solve people constantly alleging discrimination—we'll investigate and if it exists we'll take care of it."[24] Rankin's memo of the previous year explicitly noted that individual acts of discrimination had taken place.

What of the couple mentioned at the beginning of this book? The episode of the African student and the white girl described in the student newspaper occurred a year before the incidents recorded in the first chapter. The girl in the wheelchair was Lucy Correnti, and she still lives in Pittsburgh.[25] When I asked her about the story that had circulated in the sixties about a white girl

23. Letter from Rankin to Litchfield, November 2, 1962, box 129, folder 1060, Litchfield Papers, University of Pittsburgh Archives.

24. "New Board to Hear Alleged Bias Cases," *Pitt News*, October 28, 1963, 1.

25. The following section is based on my telephone conversations in August 2006 with Lucy Correnti, Jim Spruill, and Allen Janis.

in a wheelchair who had a relationship with a black blind boy, and the rumor that the dean of women had broken them up, she said, "Yes, that was us, and much of what you said is true." But she was able to fill in the details and correct parts of the story.

They had not "disappeared," as many had thought at the time. They had continued their relationship, and in 1966, they married and moved to Squirrel Hill, a largely Jewish neighborhood not far from the university. Her husband, Jim Spruill, graduated from Pitt in physics, went to Duquesne University Law School, and recently retired from a career teaching criminal justice at the Community College of Allegheny County. She went on to complete a master's degree in social work and is still active in what has been a long and productive career as an advocate for the disabled. They have two children and four grandchildren. Although their marriage ended at the beginning of the 1980s, they continue to be in amiable contact.

The dean and assistant dean of women had tried to break the couple up. Both Lucy and Jim had known before then that they would be wise to avoid the attention of the Dean of Women's Office, and for a time they succeeded. They had tried to stay under the radar and appear as just good friends; perhaps their disabilities served as a cover. But, ultimately, the dean's office found out. The dean then tried to have Lucy kicked out of the dorm. Given the difficulty of trying to live in Pittsburgh as a single handicapped woman, such action, had it succeeded, might well have ended her college career. Even worse, the assistant dean contacted the Office of Vocational Rehabilitation and informed the counselor there that Lucy intended to get married and that marriage wasn't a "vocation." Lucy had been receiving her tuition and expenses from a program administered by the Office of Vocational Rehabilitation, and since she came from modest means, had her funding been cut off, she would have had to leave school. The counselor contacted Lucy and told her that they might not continue supporting her. Fortunately, Lucy remembered that a counselor of hers when she had been a resident in the Home for Crippled Children in Western Pennsylvania had promised that he would help her in the future if the need arose. She visited him to see if he could be of some aid, and he must have been an experienced advocate, because he immediately called the supervisor of the counselor at Pitt and had the issue resolved.

Although Lucy resented the actions of the Dean of Women's Office, she appreciated the support she received from others at Pitt. Jim recalls that he found faculty supportive. Raised in Philadelphia, Jim had been blind since the age of one and a half and had attended the Overbrook School for the Blind there before coming to Pittsburgh. Although retired now, he has continued to teach part time. He remembers that there had been "massive support" for them in the Physics Department, and that the attitude of the administration had shifted when the department threatened to leave en masse. Jim had been told by his adviser that the faculty had threatened the administration that it would be packed and on its way somewhere else if the pressure on Lucy was not dropped. This was in the good old days of the sixties, when university scientists (and even some social scientists and humanists) were in demand, and faculty were more mobile than they are today. And it was a time when students and faculty put their careers on the line for idealistic causes, particularly for civil rights issues.

The essence of Jim's memory, that faculty had protested the administration's interference in the lives of Lucy and Jim, is corroborated by his former mentor in the Physics Department, Allen Janis (still active at Pitt). At the time he was an assistant professor. He and a senior member of the department (professor Thomas Donahue) had spoken to Assistant Dean Skewis. They indicated to her that many of the faculty in the department were outraged by the pressure being applied to Lucy, and that they had inferred such actions were motivated by Lucy's relationship to Jim. Janis recalls that a number of the faculty felt very strongly and that, indeed, they might have resigned had the issue not been dropped by the Dean of Women's Office. He counts himself in that group, and he thinks he might have said something to Jim to the effect that some physics faculty were so outraged by the Dean of Women's Office that they were on the point of resigning. There had not been a meeting of the faculty to discuss the issue, or a resolution to quit en masse. But there was faculty support. A faculty member in the law school (a person in an interracial marriage) was also helpful. In the end, then, enough pressure had been put on the administration that the attempt to break up the couple ceased.

Jim did not experience as much pressure as Lucy did. The dean of men, in contrast to the dean of women, was pretty liberal and treated Jim well. But if Jim felt less hassled than Lucy, he did experience some "peculiar" incidents. For example, one evening when he and Lucy were in a reader's room in the student union (the two of them set up the first reader service at Pitt for blind students), a guard came down the hall locking doors. The union had a distinctive locking mechanism in its doors. From the inside, one could put in the key and lock the door, and when that was done, it could not be unlocked (or locked) from the outside. Similarly, when locked from the outside, someone inside the room could not unlock the door from the inside. Jim and Lucy heard the guard lock their door, and they immediately realized that unless they caught his attention, they were going to be in a difficult situation. Students were not to lock a student union door when they were in one. They yelled out, and he came back (although he was not happy about the situation). A few days later, Jim was called into the Dean of Men's Office. The dean told him, "You know this is against the rules," and when Jim explained that it was quite impossible that he and Lucy had locked the door, the dean just repeated his somewhat severe admonishment. Jim was puzzled about the dean's denseness and wondered why the dean was being so strange. Was it some threatening posture that the administration had suggested he pass along? Jim doesn't know, but the interview struck him as odd, and he still is not sure how to interpret what was happening.

The initial years for this interracial couple in Pittsburgh were difficult. They especially had trouble renting an apartment. When landlords found out they were of mixed race, they refused to rent to them, and finally Lucy had to resort to looking for an apartment with a (white) friend from a civil rights group. Once she had a lease for the apartment, the landlord had little choice but to continue renting to them because the antidiscrimination laws of the time made it difficult to evict a couple based on race. Ultimately, they were happy living in Squirrel Hill, and their children, for the most part, found growing up there a positive experience. Nowadays, mixed-race couples are hardly noticed there, but that is forty years later.

Lucy remembers another interracial couple that was harassed by the administration. She heard that the dean of women had contacted the parents and that the parents had removed the girl from school (she enrolled in a small school in her hometown). There are, however, no documents in the university archives that verify the action (but then again, there was nothing in the archive about the pressure put on Lucy).

One has to admire the strength of young people like Lucy and Jim, who were able to withstand the strong social pressure against interracial marriage. Although the scientific racism that had justified earlier opposition to miscegenation had been intellectually undercut in many ways by the new genetics and anthropology, social mores change slowly and are influenced by many factors. Lucy had been raised in a middle-class family, typical of the fifties. She had absorbed Christian values and had been taught to obey the rules, respect the law, and defer to authority. It was difficult, then, to be doing something that seemed perfectly normal (falling in love) and to incur so much disapproval. The experience of going against authority (in the persons of the dean and assistant dean of women), breaking with her family for over a decade, and having to resort to subterfuge to rent an apartment went against much of her upbringing. But it taught her that she could be "empowered." She could challenge the power structure, and she could make the kind of life she wanted for herself and others. It was an important lesson, at the very heart of the sixties rebellion, and one that remained with her the rest of her life. She has been active in civil rights causes. Professionally she became the Americans with Disabilities Act coordinator for the city of Pittsburgh and currently is the director for community living and support services in the Attendant Care Program, a division of the United Cerebral Palsy of Pittsburgh that provides adults with home- and community-based services to help with consumer choice and to help them avoid institutional placement for long-term care. Her life has been one of service, and many thousands of people have benefited from her skills in advocacy.

Another point worth noting about the incident at the University of Pittsburgh is that the "administration" was not monolithic. Yes, the dean of women and the assistant dean interfered in the lives of interracial couples, but other members of the administration worked to reduce discrimination and fought to neutralize the acts of people like Helen Pool Rush and Savina Skewis. Litchfield and Rankin moved the university in positive directions and put pressure on the community to create a more open and tolerant society. Individual faculty made a difference in providing support, and individual administrators contributed to change. If Helen Pool Rush was keeping a list of who went to Selma, others were publicly praising those who went. That was a lesson that was somewhat lost on many of the sixties generation (at least until they passed thirty themselves).

And finally, the Pittsburgh story demonstrates how the issue of interracial mixing reflects the overall trends of the sixties. It was the nexus of the civil rights movement, the liberation of students from an *in loco parentis* philosophy of university administration, and the sexual revolution that broke with the conventionality of the fifties. It was an answer to writers such as Gunnar Myrdal that American society had not followed through on its promise of full recognition to black citizens.[26]

Jim and Lucy's problems, then, were not unique. Racial problems at the University of Pittsburgh mirrored the problems at other universities and were typical of what many college-aged students across the country experienced.

Although interracial dating continued to be an issue in various ways on U.S. campuses, by 1966 (the year before *Loving v. Virginia*) it was somewhat less of a flash point between students and central administrations. This change resulted not from a new, enlightened policy on interracial dating, but from

26. See chapter 2, footnote 3.

a capitulation to students about *in loco parentis* policies. As *Time* magazine noted toward the end of 1966, "On almost every campus, students are either attacking *in loco parentis*—the notion that a college can govern their drinking, sleeping and partying—or happily celebrating its death."[27] With less interference in the private lives of individual students, there was a reduction in the role university administrators played in interracial dating, and consequently, a central factor in limiting race mixing on campuses was reduced. This is not to suggest that opposition to miscegenation on campus disappeared with curfews. But by the mid-sixties, campus culture had shifted, and students were more able to follow their individual desires—constrained, of course, by the many social pressures that worked on them from their family, friends, coaches, and campus leaders.

Race mixing continued to be a threat to some whites who wanted to keep blacks from joining the increasingly mixed, multiethnic, but not color-blind society. And increasingly, the black community was ambivalent or hostile to those who fraternized with the other side. In the late sixties, the Black Power movement was hostile to interracial dating and marriage, and to this day the topic is a vexed one in the African American community. Black women in particular were skeptical about cross-race dating and marriage (even on well-integrated campuses today) in part because a larger percentage of mixed couples are black males and white females. The past exploitation of black women by white men was hard to forget, and the perceived scarcity of suitable black males available for them to marry lent further support to their position: it reduced an already limited pool of potential husbands.

Powerful social forces, then, kept interracial mixing relatively low. Opportunities like higher numbers of black students on formerly white campuses (and particularly the presence of black athletes) worked to increase the number of mixed-race couples, but the numbers of minority students remained relatively small, and so the social playing field never became leveled. Even in "liberal" families, the realities of social prejudice severely dampened an enthusiasm for race mixing. It was an issue that spanned generations and demonstrated the residual effects of racism in the United States.

27. "Universities," *Time*, 88, no. 21 (November 18, 1966): 95.

7 Science, "Race," and "Race Mixing" Today

America has changed dramatically since *Loving v. Virginia*, and part of that transformation has increased the number of, and changed the attitude toward, interracial couples. I have not attempted to describe or analyze this long and complex story. Rather, this book is intended to contribute two points to understanding the background of this transformation: the role of changes in scientific theory that undercut the scientific arguments against miscegenation, and the tension that this knowledge created, as reflected in discussions on campuses about interracial dating.

This final chapter will consider some other issues that relate to shifting opinion today in the scientific community on race and race mixing. First, I will discuss how different scientific disciplines view race (and by extension, race mixing) and how disciplinary perspectives are sometimes in conflict. There is not a single "scientific" interpretation of what *race* means currently. Second, I consider the broader issue of attempts to apply biological knowledge to gain insight or guide social values. Since Darwin's day, writers have sought lessons from biology that they hoped could be applied in the human realm. The record of success has not been encouraging, and this in itself may hold an important lesson. In particular, I want to examine ambiguities within the modern theory of evolution and its implications for our understanding of race and of race mixing.

Varieties of Interpretation of Race

The modern theory of evolution has been effective in combating racism. For example, if one looks at recent attempts to revive forms of blatant scientific racism by such people as Philippe Rushton, or Charles Murry and Richard Herrnstein, we see strong negative reactions within the scientific community. Nonetheless, an ambiguous legacy stems from the modern synthesis. It was foreshadowed in the debate between Dobzhansky and Montagu over the use of the term *race*. Dobzhansky held that races exist in nature: they are breeding populations that can be distinguished by their differing gene frequencies. How we designate them depends on our purpose, and the resulting classification will vary with the characteristics that we choose. Everyday folk categories, such as Black or Asian have little or no biological meaning, but that isn't to say that we cannot construct meaningful biological groups for particular purposes (e.g., Ashkenazi Jews, for medical classification). Montagu rejected Dobzhansky's position, and he was famous for stressing not only that *race* is a social construction, but also that we should purge the word from our vocabulary because of its toxic associations. Their disagreement has taken on a life of its own, but it has not been especially evident because those who hold different positions in this debate are often not in direct contact.

People in different disciplines sometimes define basic concepts differently, occasionally so differently that they are fundamentally in conflict, or simply inconsistent. In a typical Department of Zoology, for example, faculty find it normal to talk about the race, variety, or subspecies of a bird or insect. These terms generally refer to a geographical group of the same species that inhabit different portions of a continent, or live on different continents. From an evolutionary standpoint, they reflect the process of speciation; they are species "in the making." Large battles have been waged over how to classify them: should they have separate names, or not, for example. The concept goes back to Darwin's time (and earlier) and is part of the intellectual apparatus with which zoologists (and other naturalists) order the living world.

If one walks across campus to the Anthropology Department, quite another opinion reigns. There, the faculty generally consider race a social construction, one *without* a biological meaning. In fact, they avoid the term or, if used, carefully qualify it. Montagu, or one of his followers, such as C. Loring Brace, is likely to be cited, along with the contemporary Harvard biologist Richard Lewontin. Lewontin wrote an influential paper in the early seventies in which he argued based on his study of proteins that greater variety exists within a race than between or among races. Anthropologists have used this information to bolster Montagu's position that race represents a deeply problematic concept.

Zoologists, of course, are well aware of the extraordinary variation within a species—a point Dobzhansky often stressed, and one that has been recognized in systematics for over a century. Animals of the same species vary significantly, and the difference between races is not as great as the overall variation within the species. But zoologists nonetheless find it useful to distinguish populations with relatively small differences. Dobzhansky tracked how minute differences in subpopulations of *Drosophila* change over space and time, and argued that they were the keys in understanding the emergence of new subspecies and new species. So, although no one doubts Lewontin's point, it does not negate the value of using the concept of race in zoology. Many anthropologists feel otherwise, at least when it comes to humans. Although medical and forensic anthropologists identify subpopulations of humans and use them in their work, the greater number of anthropologists are keenly aware of the history of scientific racism (to which their field contributed so much) and want to distance themselves as far as possible from anything that could resemble it. The debate between Montagu and Dobzhansky, therefore, remains alive on campuses, and students can get quite different messages from different instructors and different texts.

If one were to query professors in history or philosophy, they would mostly agree with the anthropologists. Historians argue that the concept of race has changed significantly over time, and that the term has been used in multiple ways. They focus, of course, on how the term has been used about humans over the past century or more and concern themselves less with, for example, how it may have been used by ornithologists in their attempts to order birds of New Guinea. They therefore regard *race* as a constructed term, one deeply conditioned by history.

These different perspectives on race sometimes emerge in controversies, for example, the heated debate about Kennewick Man. In 1996, some college students uncovered along the Columbia River in Kennewick, Washington, a nearly complete skeleton, which turned out to be more than nine thousand years old. The Kennewick Man controversy has lasted for over a decade. The root issue stems from a disagreement between a group of scientists, who want to study this potentially interesting skeleton, and local Indians, who claim it as an ancestor who should be respectfully buried and not desecrated by the probing fingers of anthropologists or their instruments. In 2004, the Ninth U.S. Circuit Court of Appeals ruled in favor of the scientists.

Part of the conflict had to do with the very poor way in which the initial discovery was handled (the tribes were not notified immediately, etc.). But the discussion quickly became one about race. The first anthropologist to see the skeleton, James Chatters, thought the skull belonged to a Caucasoid male; that is, it was not the remains of a Native American. The claim had considerable importance because ideas about the early populating of the New World were (and are) being revised. Because of the discoveries of anomalously early human remains in the Americas, many anthropologists have come to doubt the conventional view of how the humans reached the Americas (by crossing a land bridge from Siberia about 13,500 years ago). Also, the study of some early North American skulls showed that these early remains appeared to differ

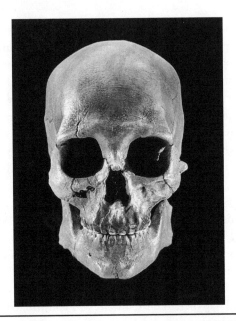

Race in Anthropology

Kennewick Man, shown here, now resides in the Burke Museum of Natural History and Culture at the University of Washington. The discovery of this skeleton in 1996 sparked a lively controversy among anthropologists, Native Americans, and museum curators. Race was at the heart of the debate since an independent archaeologist who first studied the bones claimed they were the remains of a human who belonged to a different racial type than Native Americans currently living in the Northwest.

■ Reconstruction of Kennewick Man skull. Photo by Chip Clark. Courtesy of Smithsonian Institution.

significantly from the skulls of more "recent" American Indians. Many anthropologists, therefore, have concluded that there were multiple migrations, some perhaps from Europe. Kennewick Man, especially since he is so complete (and has a spear point embedded in his hip bone), holds considerable interest.

So how does race factor in? Partly, the use of the term "Caucasoid" by Chatters and others suggested that Kennewick Man belonged to a different racial group than modern Northwest Indians. Forensic anthropologists have no trouble using racial terms, but as noted, most of the anthropology profession has dropped references to race as a meaningful scientific term. For example, members of the Department of Anthropology at Oregon State University, with the exception of the one person involved in the lawsuit to gain access to study the ancient skeleton, sided with the Indian tribes. The argument they advanced contended that since "racial" groups contain vast variation, there is no reason to assume that Kennewick Man belonged to a racial group unrelated to contemporary Northwest Indians. He might have been taller, had a longer face, and so on, but within the Indian population, one could find all manner of variation. They also noted that since Indian oral traditions claim to extend ten thousand years, we should respect their claim of Indian ancestry for Kennewick Man. Rob Bonnichsen, the dissenting voice in anthropology and the director of the Center for the Study of the First Americans at the time, wanted to examine Kennewick Man to extend the use of mtDNA studies for unraveling the complex history of the populating of the Americas. In part, he became interested in the subject when he uncovered some strands of hair in an ancient site in Russia and was able to extract some mtDNA from it. He combined this with traditional studies of skulls in an attempt to help sort out the record.

Studying skulls, of course, reminded his colleagues of the bad old days of physical anthropology, when scientists tried to construct typical profiles of racial groups. Bonnichsen was not trying to do that. (He died in 2004, so he was not around to examine the skeleton.) Rather, he was trying to piece together a complicated story using different bits of evidence and with a full recognition that populations change in time, mix, and are exceedingly difficult to characterize. The number of skeletons, or partial skeletons, available remains very small, and therefore any such studies will always be on the margin of speculation, as are practically all the attempts to construct fossil histories of modern humans. What he wanted to do resembles what a zoologist would do when presented with a specimen of a bird and asked to identify it. Depending on the species, the zoologist might be able to identify it all the way down to a particular geographical subspecies (race). In so doing, of course, he would realize that each subspecies has great variation, and that it is perfectly possible that the individual might come from an entirely different population. If the zoologist were then presented with the interesting information that the bird had been found in an area outside its normal range, he might wonder about that and reconsider. The bird could be a variation, or the ranges of the subspecies might have to be revised. More information would be needed. And of course, that was Bonnichsen's point: that we need to study every skeleton that we can find if we want to understand how many migrations there were, and where they came from.

There are, to be sure, serious issues surrounding the conflict between the desire of anthropologists to study the history of Native American populations and the unwillingness of contemporary members of the Native groups to desecrate the graves of their ancestors to uncover information. But it is interesting to note how quickly within the Oregon State Anthropology Department the issue of race intruded on the conversation over Kennewick Man and muddied the water. Five members of the department wrote a letter to the local newspaper, distancing themselves from Bonnichsen and stating that scientific analysis of skeletal remains "cannot and should not be used to further theories about so-called 'paleo-racial' differences between populations in the Northwest."[1] So, the debate between Montagu and Dobzhansky has not been settled. Could it be? I suspect so. Dobzhansky knew that popular and folk classifications had no biological meaning, and Montagu knew that Mendelian populations of humans existed. Scientists construct classification systems for particular purposes. If it turns out that certain individuals who have family ties to particular geographical areas that in the past experienced some isolation, then those individuals might have a greater (or lesser) frequency of some genes, and that information might be useful to know for scientific or medical reasons. Currently, physicians and medical ethicists are debating this issue, even as pharmaceutical houses are moving ahead with drugs that target particular "racial" groups, that is, groups of individuals who share some ancestry, be it West African or eastern European. One has to wonder how successful the pharmaceutical company strategies will be, given the small number of genetically associated diseases (e.g., sickle cell or Tay-Sachs) compared with the number of likely environmentally associated diseases that disproportionately affect individuals of different racial and ethnic groups. African Americans experience more problems with high blood pressure, for example, than the general population. Although there may be unknown genes involved in a predisposition to high blood pressure, there are so many known environmental factors (diet, stress, etc.) that concentrating on a genetic fix may be counterproductive on a national public health level, especially if it channels time and money away from attacking the social and economic issues that cause individuals to develop high blood pressure.

So the message from the modern synthesis remains ambiguous, or at least, its interpretation remains so. We may be able to identify subpopulations of humans, but what we do with this information remains problematic. Anthropologists, who in the United States had a long and shameful association with grave robbing Native remains, and who in the past ranked "races" in a hier-

1. Joan Gross, Sunil Khanna, Deanna Kingston, David McMurray, and Barbara Roth, "Scientists Respect the Tribes' Views," *Mid-Valley Sunday Corvallis Gazette Times*, July 1, 2001, A5.

archy, are understandably concerned that we not return to the old scientific racism of the past century. Zoologists, on the other hand, feel their field was less involved in scientific racism, and that indeed the study of variation in subpopulations helped destroy the faulty views on racial hierarchies. For many of them, the notion that "the truth will set you free" has greater resonance. At a minimum, we need to be sensitive to the underlying assumptions, agendas, and contexts of the discourse. It was one thing in the 1930s for Nazi supporters to write about Jews as a race, but quite another to talk today about a subpopulation of Europeans, Ashkenazi Jews, who by largely marrying within their community created a Mendelian population that differs in the frequency of a few genes from neighboring populations, and therefore are at a higher risk for some genetic diseases.

The construction of race may turn out to be an activity that relies on social cues—language, dress, class—but also contains elements of biological reality—skin color, BRCA-1 genes. Although the variety within any racial group exceeds the variety between it and another, it takes a strong commitment to abstract principles not to see that, on average, the people in Sweden appear different from the people in the Congo, or that some diseases seem to strike more individuals of a particular ethnic background than others. It may be a minor difference that distinguishes groups (skin color, or a gene that is not visible), but it informs how people think about the subject nonetheless. Despite modern biology, people find it difficult to bear in mind that the traits of "races" characterize groups (frequencies of various genes), not the individuals who compose the groups. What we encounter in daily life are individuals, and, perhaps more important, they are generally what matter.

Mixed-race couples, and their children, confuse the story even more for some people. That is, race mixing threatens the notion that human groups have inherent differences. It calls attention to the deeply mixed nature of all human groups and subdivisions of groups. It also uncomfortably reminds people of the social construction of most human groups. Further, it undercuts the ideology of difference or the belief in the essential difference among closely related groups: Serbs/Croats, Hutu/Tutsi, Palestinians/Jews, and so forth. Mixed-race couples, however, also point to a future where these distinctions may seem irrelevant. In the 2008 election campaign that Barack Obama won, the issue of mixed race was prevalent. Pictures of Obama and the white grandparents who raised him were interspersed with shots of him on the South Side of Chicago surrounded by blacks. His accent recalled the elite institutions where he went to school and where he had taught, but his rhetorical style reminded many of the power of black orators, like the famous Martin Luther King. Barack

Obama is the first African American president of the United States, but he is also the first clearly mixed-race president of the United States. And as such, he overturns many of the stereotypes that plagued "mulattoes" in the twentieth century.

Evolution and Social Values

There is another, and deeper, sense in which the ambiguity of interpreting the modern synthesis appears problematic: the use of evolution and genetics to justify social positions or policy (as opposed to scientific ones). The issue can be seen in Dobzhansky's work. Its high moral tone strikes anyone reading his writings. He typifies the scientist who sees the big picture, not only in science, but in questions about our basic humanity. In 1967, for example, the same year as *Loving v. Virginia*, Dobzhansky published *The Biology of Ultimate Concern*, a discussion of philosophy and religion in the book series Perspectives in Humanism. Even reading Dobzhansky's papers on *Drosophila* that relate to subspecies formation, mixing, and evolution, one sees that Dobzhansky wrote explicitly about the human implications of his genetics research. He claimed that evolution holds lessons for humankind, among them that race mixing is a normal activity that maintains the strength and diversity of the gene pool. He believed greater diversity represents a positive good, and stressed the point in numerous scientific and popular writings. His position, however, occasionally led him to awkward situations. For example, although throughout his career he combated racism, he held a somewhat tolerant view of Arthur Jensen's writings that hereditary factors account for some of the difference between white and black IQ scores. Dobzhansky could imagine that genetics might account for some differences in intellectual performance. And if that were so, it might be that some populations differed intellectually. Jensen extrapolated from his IQ studies to suggest that programs like Head Start, which attempted to boost the academic achievement of African American (and other) students, were doomed to fail because of the lack of inherent intellectual ability of some groups. Dobzhansky strongly rejected such conclusions, maintaining that the degree of variation within groups remained significant, but also recognizing the importance of environmental influences on behavior. Nonetheless, he could not go along with the Jensen's critics, who wanted to deny any degree of heritability of skill or behavior.[2] Since Jensen's writings attracted the charge of being racist, Dobzhansky's position could be seen as lending partial support to a view that contradicted his decades-long battle to combat racism.

2. See Diane Paul, "Dobzhansky in the 'Nature-Nurture' Debate," in Mark Adams, ed., *The Evolution of Theodosius Dobzhansky* (Princeton, N.J.: Princeton University Press, 1994), 219–31.

The point is that arguing from biological theories to justify social policy or social views has long been a perilous strategy. As we have seen, the eugenics movement of the early part of the century read social values into nature and attempted to legitimate policy based on them. The strong negative reaction against eugenics has not, however, diminished the attraction of using evolution as a possible worldview and inferring conclusions from it. The fields of evolutionary psychology and evolutionary ethics have suggested to numerous scholars, and popular writers, ways of using the theory of evolution to explain aspects of human nature, everything from justifying moral standards to understanding dating preferences. Books and articles have extended these "insights" to argue for a wide range of social positions and policies. An evolutionary perspective, for example, allegedly illuminates a number of topics discussed in popular magazines like *Psychology Today*: restricting immigration, tolerance for gays and lesbians, promoting ethnic pride, and changing laws that restrict the legal age for marriage.

In this study, we have seen how in the early part of the century, Davenport, an important scientist at the time and one who thought he was separating science from prejudice, in fact read his own racial views into his science. He was convinced that race mixing held dangers, and in spite of the basically useless data from Jamaica, he continued to espouse that opinion. His concern about the mixing of whites and blacks in the United States was couched in a similar claim of "objectivity" and concern for preventing matches that would result in less fit offspring. But he had no solid database with which to support those views.

Franz Boas frankly admitted that the empirical record was weak on the issue of race mixing and honestly confessed that his views could not be supported by much hard data. What about Dobzhansky? He was certainly on firm ground when he noted that the Indian caste system had not produced human isolates that displayed decreased fertility, and that most of the arguments against race mixing were invalidated by the modern theory of evolution. And yet he dismissed rather cavalierly the notion that crosses of individuals from widely divergent populations with possibly different adaptive gene complexes might produce individuals who are less fit. Why could he dismiss this so easily when Davenport and others had firmly believed it? Was Dobzhansky's science just as deeply driven ideologically as Davenport's?

When one reads Davenport, Mjoen, and East's comments on race mixing, it doesn't take much to determine how shaky their scientific justifications were. Dobzhansky, and the other architects of the modern synthesis, undercut the scientific racism of the earlier part of that century by shifting to population thinking and revising the concept of species. And modern physical anthropology has decidedly rejected folk classifications that have no biological meaning. But what about Dobzhansky's flat rejection of the possibility that some human populations might contain gene complexes that had diverged in time significantly from other populations? Did Dobzhansky allow his social perspectives to influence his biological ones? It wouldn't be unusual, but it would undercut the high moral ground that he had staked out.

Can scientists rise above the social context in which they work? Most historians of science would argue that science has social values deeply embedded in its conceptual framework, and that seeing these assumptions is, if not impossible, at least very difficult. James Watson, one of the discoverers of the structure of DNA and an eminent scientist, revealed how easy it is to mix one's own values and biases with lessons drawn from nature. In an October 14, 2007, interview with the *Sunday Times* (London), he suggested that black people are less intelligent than white people. The paper quoted him as saying that he remained pessimistic about the future of Africa because social policies are based on the assumption that the intelligence of blacks was the same as whites, but that testing suggests otherwise. He compounded the problem by adding that although he hopes everyone is equal, those who deal with blacks as employees discover otherwise. Watson's statement, although it appeals to observation, does so in a manner wholly different from the experimental observations (and interpretations) that led him and Francis Crick to propose their model of DNA. Rather, in his comments on race, Watson was relying on his own social perceptions and biases. His public gaffe led to his resignation (formally, his retirement) as chancellor of the Cold Spring Harbor Laboratory (coincidentally, Davenport's former laboratory). That such an eminent scientist could make such a blatant social blunder, however, has to give us pause.

History has a didactic value that can make us sensitive to the misuse of concepts. It may not be able to shield scientists and the public from all error, but it can serve as a cautionary tale. Further, it should incline us to look carefully at claims of bias in science and be willing to examine how discrimination may be built into our institutions. For example, feminist scholars have long argued that the contemporary scientific enterprise remains structured so that it inherently favors men. If we can see this, we can change it, and efforts have been made in recent years to address some aspects of this problem. The history of the changing scientific attitudes toward race mixing clearly demonstrates the degree to which social bias can penetrate science. It also shows how individuals, and groups of individuals, can move beyond the prejudices of the past and use science to unmask and eliminate some of these biases.

Epilogue

This book began with a story from the sixties about a mixed-race couple at the University of Pittsburgh. Many of their fellow students were outraged by what they heard of interference by the dean of women into these two students' lives. The sixties generation wanted change, and they wanted it *now*. But change comes slowly. It has been a lesson that the sixties generation was loath to accept but that experience and historical studies of civic institutions support. It would be nice if *Loving v. Virginia* had ended the story of opposition to race mixing in the United States. But it did not. A change in law often brings about a change in society, or it can reflect a changing attitude, but the process is slow.

The university where I taught for most of my career, Oregon State University, is located in Corvallis, a small, predominately white college town. The school dominates the city and contributes to its middle-class atmosphere. I moved there in 1970, just three years after *Loving v. Virginia*. On the surface, the mood, as reflected by voting patterns and local government, is moderately liberal. But the state has a mixed record in race relations. The Ku Klux Klan maintained a strong presence in the early part of the twentieth century, and Portland, the largest city in the state, did not have a single hotel that would accept a black client in the fifties.

At Oregon State, black athletes had a difficult time in the sixties, as they did in many schools across the country. A standoff between black students and the administration occurred when a black football player was told by his coach that he had to shave his beard. The resulting demonstration represents just one of many similar disturbances on other college campuses involving the increased number of black students in what had been predominantly white schools.

A colleague told me a few years ago that Amory "Slats" Gill, the basketball coach (1928–1964) after whom the coliseum is named, refused to recruit black basketball players because he did not want them dating white girls, and there were virtually no black coeds. The coach would have gotten along just fine with the dean of women at the University of Pittsburgh, as well as with many other university leaders across the country. But that describes a world before the impact of the civil rights movement. There have been only a few racial incidents on campus during my time at OSU, and the university administration has made several moves to reduce future causes of others. "Campus climate" surveys, however, indicate that people of color still feel they are less welcome than others. Although interpreting those studies presents serious difficulties— they were not conducted in a systematic manner—they nonetheless suggest a continuing set of issues involving race and ethnicity on campus.

A few years ago, a story appeared in my local newspaper about a young married couple celebrating at a bar/club the last night of the national guardsman's leave before he returned to Iraq. The couple was mixed-race, white husband, black wife. A few underage black athletes entered the bar and began harassing the couple. Words were exchanged, and upon leaving, one of the athletes hit the guardsman and knocked him out. Much could be unpacked from this story (Why were underage athletes in a bar? Why were they involved in barroom brawls? etc.), but what caught my eye in the paper was the reference to how the young athletes harassed the black woman about being with a white man. Maybe in Pittsburgh a mixed-race couple can go relatively unnoticed, but in much of small-town America, less so.

I wish I had a solution. This study of changing attitudes toward race mixing suggests that change has occurred, but at a slow rate. We need to keep in mind that change can happen, and has happened. Presently, at Oregon State (and at many others like it), "diversity" has emerged as a major institutional initiative. All searches—administrative, faculty, staff—have to address the issue. The administration has broadly defined diversity to include many categories of how people differ (skin color, ethnic background, sexual orientation, disabilities, socioeconomic background, etc.), and the campus has a goal to create a positive attitude toward diversity ("celebrate diversity" is the language). It's a daunting challenge. One can change language and mandate courses that educate students on the positive value of diversity and the sad story of discrimination in the United States, but people change their preconceptions very slowly, and in an institution like the university, the population of students changes from year to year. New freshmen arrive each year with backgrounds over which the university has little control. Family attitudes, high school experiences, peer perceptions all influence how new students see their fellow students. Compounding the problem, groups of similar students often find they are comfortable together, and so a ghettoization of campuses commonly occurs, even in the most liberal and progressive of schools.

That official attitudes have changed should make a difference. A mixed-race couple today would not experience the official censure of university administrators, and classmates who listen to what orientations present will hear a different message than was conveyed at the beginning of the last century. To be sure, orientations can strike students as tedious, and many of the mundane aspects of the diversity initiative threaten to undermine the seriousness of the matter. But even with the administrative overkill, it is clear that the university climate has changed concerning racial issues. It has taken time, and still has a long way to go, but there has been progress. The story of Lucy and Jim should remind us that mixed-race couples have thrived throughout this changing environment. Their children are successful. And although Lucy's family was initially hostile to the couple, after Jim became an attorney, their stance softened. Class trumped race to some degree. Lucy and Jim were contemporaries of Barack Obama's mother and father. Both families broke up, Barack's a good deal earlier than Lucy and Jim's. But, Barack's career shows a degree of acceptance of a mixed-race child on a national scale that was unimaginable a few decades ago.

The story of ideas on race mixing in the period 1937–1967 shows a major shift in attitude in the United States. In the thirties, many states had anti-miscegenation laws, and the medical community believed race mixing was potentially deleterious. By the time of the Supreme Court decision in 1967, no respectable scientific opposition to race mixing existed, and there were more interracial couples. But it would be a mistake to interpret the story in too positive a light; opposition, social rather than scientific, continued to be widespread and practiced by members of all "races." The strife of the civil rights era convinced many blacks that their destiny was separate from whites, and racial discrimination did not disappear. So the story did not end with *Loving v. Virginia*.

The situation remains sufficiently complex that it may not be possible in our lifetime to put it all behind us. There exist no genetic barriers to race mixing. Blacks and whites in the United States are socially constructed groups that have little or no biological meaning. Medically and forensically, there may be distinctions that are not spurious, but so much mixing has occurred that we need to be extremely cautious in how we use "precision" instruments such as genetic markers in defining groups. They tell us very little that relates to the broader subject of race mixing. But what of the social barriers? Biological barriers may not exist, but our society splits itself into groups that desire to maintain their identities. Group identity has value and importance. A lot of recent politics has been identity politics, and many biologists would argue that as social animals, we have a deep-seated psychological predisposition to identify with groups. Whether or not we accept that position, history shows that humans have fierce loyalty to perceived groups and have been willing to die to maintain borders. One of the main themes of European history since the early modern period has largely been the rise of nationalism and the clash of groups of individuals who believed they belonged to different nations. The late twentieth century, unfortunately, has not seen any let-up, with ethnic cleansing becoming a lamentable part of the historical description of events on all five continents. Mixed couples in many such regions have been among the collaterally damaged.

People seem to want boundaries. Ethnic pride, nationalism, and racial identity can motivate individuals to creative heights. But these same boundaries can harbor discrimination and can be destructive. The challenge for the coming generation is to construct a dialogue to determine where we want boundaries, where we want to dismantle them, and why. The history of ideas on miscegenation shows us how complex these discussions can be, and how damaging they can become when we get it wrong. The value of these studies does not reside in reading history to assign blame and castigate those who made the lives of others miserable, but in taking lessons for the future. Human societies construct categories, and we have to be willing to rethink them, even the most deeply held ones. Just as important, we need to see how easily we confuse categories. So much of the discussion of race was, and is, actually about class or socioeconomic status. So much of the discussion of race was, and is, about groups maintaining dominance. So much of the discussion of race was, and is, about categories and not individuals. If we can understand that, then maybe we can begin to rethink what sort of society we want to construct, what obstacles we will need to remove, and what methods we need to employ to get there. As numerous commentators have suggested, perhaps we need to consider how to permit individuals to have multiple identities, or nested sets of identities, to encompass their complex allegiances. If we don't make progress in thinking through the discussion of race that has bedeviled the twentieth century, then we will continue to endure ever new stories of ethnic and racial conflict. Can we make progress? Given our ever increasing ability to inflict harm on others, we cannot afford not to.

Suggested Further Reading

For a general introduction to the history of ideas on race mixing in twentieth-century United States, see Werner Sollors, *Neither Black Nor White Yet Both* (Oxford: Oxford University Press, 1997); Paul Spickard, *Mixed Blood: Intermarriage and Ethnic Identity in Twentieth-Century America* (Madison: University of Wisconsin Press, 1989); Joel Williamson, *New People: Miscegenation and Mulattoes in the United States* (New York: Free Press, 1980); Joseph Washington, Jr., *Marriage in Black and White* (Boston: Beacon Press, 1970); Rachel Moran, *Interracial Intimacy: The Regulation of Race and Romance* (Chicago: University of Chicago Press, 2001); David Hollinger, "Amalgamation and Hypodescent: The Question of Ethnoracial Mixture in the History of the United States," *American Historical Review* 108, no. 5 (2003): 1363–90; and Renee Romano, *Race Mixing: Black-White Marriage in Postwar America* (Cambridge, Mass.: Harvard University Press, 2003).

The legal issues and background to attitudes on race mixing are well discussed by Randall Kennedy, *Interracial Intimacies: Sex, Marriage, Identity, and Adoption* (New York: Pantheon, 2003), and Peggy Pascoe, *What Comes Naturally: Miscegenation Law and the Making of Race in America* (New York: Oxford University Press, 2009). Also see Paul Farber and Hamilton Cravens, eds., *Race and Science: Scientific Challenges to Racism in Modern America* (Corvallis: Oregon State University Press, 2009).

Scientific Racism and Attitudes toward Race Mixing

A good introduction to the extensive literature on this subject is provided in the following books: Elazar Barkan, *The Retreat of Scientific Racism: Changing Concepts of Race in Britain and the United States between the World Wars* (Cambridge: Cambridge University Press, 1992); Pat Shipman, *The Evolution of Racism: Human Differences and the Use and Abuse of Science* (New York: Simon and Schuster, 1994); and Audrey Smedley, *Race in North America: Origin and Evolution of a Worldview* (Boulder, Colo.: Westview Press, 1993). Also see William Provine, "Geneticists and the Biology of Race Crossing," *Science*

182 (1973): 790–96; and Joseph Graves, *The Emperor's New Clothes: Biological Theories of Race at the Millennium* (New Brunswick, N.J.: Rutgers University Press, 2001).

The Modern Theory of Evolution and Concepts of Race and Race Mixing

For background on the origin of the modern theory of evolution, see William Provine and Ernst Mayr, eds., *The Evolutionary Synthesis: Perspectives on the Unification of Biology* (Cambridge, Mass.: Harvard University Press, 1980); Vassiliki Betty Smocovitis, *Unifying Biology: The Evolutionary Synthesis and Evolutionary Biology* (Princeton, N.J.: Princeton University Press, 1996); and Joe Cain, "Rethinking the Synthesis Period in Evolutionary Studies," *Journal of the History of Biology* 42, no. 4 (2009): 621–48.

More detailed discussion of the genetics that was central to new ideas on race and race mixing can be found in William Provine, *Sewell Wright and Evolutionary Biology* (Chicago: University of Chicago Press, 1986), and Mark Adams, ed., *The Evolution of Theodosius Dobzhansky* (Princeton, N.J.: Princeton University Press, 1994). A discussion of current scientific ideas on race can be found in Nathaniel Gates, ed., *The Concept of "Race" in Natural and Social Sciences* (New York: Garland, 1997).

Index